Solar Energy Applications to Buildings and Solar Radiation Data

Solar Energy Development – Third Programme

Series Editor: W. Palz

Volume 4

Solar Energy Applications to Buildings and
Solar Radiation Data

Solar Energy Applications to Buildings and Solar Radiation Data

Proceedings of the
EC Contractors' Meeting held in
Brussels, Belgium,
1 and 2 October 1987

edited by

T. C. STEEMERS

Commission of the European Communities,
Brussels, Belgium

KLUWER ACADEMIC PUBLISHERS

DORDRECHT / BOSTON / LONDON

for the Commission of the European Communities

Library of Congress Cataloging in Publication Data

```
EC Contractors' Meeting (1987 : Brussels, Belgium)
   Solar energy applications to buildings and solar radiation data :
proceedings of the EC Contractors' Meeting held in Brussels,
Belgium, 1 and 2 October 1987 / edited by T.C. Steemers.
      p.   cm. -- (Solar energy development--third programme ; v. 4)
   Bibliography: p.
   Includes index.
   ISBN 9027727155
      1. Solar heating--Congresses.   2. Solar radiation--Congresses.
3. Solar buildings--Congresses.   I. Steemers, T. C.   II. Commission
of the European Communities.   III. Title.   IV. Series.
TH7413.E256 1987
697'.78--dc19                                                  88-2759
ISBN 90-277-2715-5                                             CIP
```

Publication arrangements by
Commission of the European Communities
Directorate-General Telecommunications, Information Industries and Innovation, Luxembourg

EUR 11430
© 1988 ECSC, EEC, EAEC, Brussels and Luxembourg

LEGAL NOTICE
Neither the Commission of the European Communities nor any person acting on behalf of the Commission is responsible for the use which might be made of the following information.

Published by Kluwer Academic Publishers,
P.O. Box 17, 3300 AA Dordrecht, The Netherlands.

Kluwer Academic Publishers incorporates the publishing programmes of
D. Reidel, Martinus Nijhoff, Dr W. Junk and MTP Press.

Sold and distributed in the U.S.A. and Canada
by Kluwer Academic Publishers,
101 Philip Drive, Norwell, MA 02061, U.S.A.

In all other countries, sold and distributed
by Kluwer Academic Publishers Group,
P.O. Box 322, 3300 AH Dordrecht, The Netherlands.

Printed in The Netherlands

Table of Contents

INTRODUCTION

This volume provides a valuable overview of the Commission's current
activities in research on solar radiation data and in the development of
solar energy applications in buildings. It contains the proceedings
of the second contractors' co ordination meeting in the third four-
year R+D programme (1985-88) in these two fields, which was held in
Brussels on 1 and 2 October 1987. The first meeting was held on
13 and 14 November 1986.^(')

The research contractors' reports presented in this book give a full
account of activities at an important stage in the programme where most
projects have completed their preparatory stages and productive
research is now fully underway. A feature of the reports is the
pattern they illustrate of working groups in which practically all
of the Member states are represented, co operating in the attainment
of common goals within a co ordinated European programme.

The meeting was attended by 60 participants and extended over two
days. Presentations were made by the contractors on the objectives
of their projects, progress made in the research, and future work
plans and priorities. Each presentation was followed by a short discussion. The meeting was also important in improving communications
between the research groups, exposing opportunities for co operation
and collaboration, and monitoring progress.

Among key achievements reported in the area 'Solar Radiation Data' is
the implementation of a network of measuring stations at eight proto-
typical sites throughout Europe. A test phase has been completed and
results are now being obtained which will contribute to improved
prediction of the technically-usable solar radiation at particular
sites on the basis of currently-available site-specific data. There
were first reports from two recently-begun contracts: project SUNSAT, in
which statistical data on available solar radiation energy in areas of
the world with sparse radiation data networks will be derived from
measurements by geostationary satellites; and EUFRAT, which will prepare

climatological data required as inputs for solar system design tools.

In 'Solar Energy Applications to Buildings' the group concerned with
solar system model development and validation described a series of
microcomputer software packages which it now has available. Considerable
progress has also been made in the development of a complete, consistent
and reliable set of test procedures for solar storage systems, and
evaluation of a first series of tests to verify new storage models has
begun.

The increased emphasis now attached to passive solar R+D because of
its economic and architectural attractions is reflected by the number
and scope of projects which relate to this technology.[2] At the largest
scale the team developing a computer-aided management system for
urban renewal which makes use of passive solar demonstrated its progress
through sophisticated computer visualisations of a high graphic quality.
The first seven in a series of fifty attractive brochures illustrating
important monitored passive solar projects in Europe, and which are
being distributed on request to thousands of European architects, were
described.

The wide-ranging concerted action which is intended to remove obstacles
to the use of daylighting in office buildings reported on the results
of its first year's endeavours, including a number of important studies,
analyses and critical reviews. Laboratory studies of human reactions
to changing thermal environments are underway which will provide a
basis for a mathematical model of thermal comfort in transient conditions, such as are experienced in passive solar buildings.

At a more detailed level, progress is being made in developing automatic
control systems for passive systems. University/industry co operation
is bearing fruit in proposals for sophisticated but adaptable optimising
controllers, while other work is developing a multi-tracer gas system
for measuring the movement of air betwen zones in buildings. Features
of the latter include the suitability of the system to rapid installation
in buildings, low capital cost, and useability while the building is
occupied.

The project PASSYS has the twin aims of increasing confidence in passive solar simulation models and design tools, and developing reliable and affordable test procedures for passive solar components.   The means are a series of standard test cells throughout Europe, and a mainframe computer simulation code.   Test cells have been commissioned and a series of sub-groups was able to report on progress in a wide range of topics, with every indication that very valuable conclusions will emerge.

Finally, the special issues which arise in ensuring that the results of this richly diverse research effort are rapidly made available to the European industry in the most effective manner were also addressed. SOLINFO aims to improve the understanding of the particular information requirements of the various building designers and constructors at the different stages of the building process.   ARCHISOL has the goal of introducing both practicing architects and students to energy efficient and passive solar design in a stylistically neutral manner.

Reference must be made to the newest project, 'Building 2000'.   This will encourage the use of passive solar techniques in the design and construction of non-domestic buildings by providing support to design teams in real building projects.   Thus a very positive link is established between the research activity and the industry.   A call for proposals launched in March 1987 attracted an enthusiastic response.   Over 70 proposals were received involving all types of buildings.   At the time of writing, final decisions are being made on the most suitable projects for support.

References

(') T C Steemers (Ed.)
    Solar Energy Applications to Buildings and Solar Radiation Data -
    proceedings of the EC Contractors' Meeting, Brussels, November 1986
    D Reidel 1987
(²) W Palz (Ed.)
    1987 European Conference on Architecture - proceedings of an
    international conference held at Munich, 6-10 April 1987
    H S Stephens & Associates

AGENDA

Contractors' Coordination Meeting - Research Areas A and D

Thursday, 1 October 1987

09.00 - 09.15   Introduction by the Commission
                 W. Palz, Head of the Solar Programme
                 T.C. Steemers, Responsible for Research Areas A and D

09.15 - 10.45   European Micro-climates

                 Coordinator: K. Scharmer, EN3S-0049-D
                 C. Villien, Sub-contract to EN3S-0049-D
                 G. Hug, Sub-contract to EN3S-0049-D
                 A. de la Casinière, Sub-contract to EN3S-0049-D
                 H.B. Spencer, Sub-contract to EN3S-0049-D
                 J.I. Jimenez, Sub-contract to EN3S-0049-D
                 L. Giacomelli, Sub-contract to EN3S-0049-D
                 A. Joukoff, Sub-contract to EN3S-0049-D
                 G. Luther, Sub-contract to EN3S-0049-D
                 D.P. Lalas, Sub-contract to EN3S-0049-D

10.30 - 10.45   Coffee Break

10.45 - 11.15   SUNSAT (Radiation Data from Satellite Images)

                 Coordinator : E. Raschke, EN3S-0112-D

11.15 - 11.45   EUFRAT (Cumulation Frequency Curves of Global Solar
                 Radiation)

                 Coordinator : J. Adnot, EN3S-0111-F

11.45 - 12.30   OPSYS:

                 Coordinator: W.L. Dutré, EN3S-0046-B
                 J. Adnot, Sub-contract to EN3S-0046-B
                 A. de Geus, Sub-contract to EN3S-0046-B
                 S. Ostergaard-Jensen, Sub-contract to EN3S-0046-B
                 C. den Ouden, Sub-contract to EN3S-0046-B

12.30 - 14.00   Lunch Break

14.00 - 15.00	Solar Storage Testing Group:

Coordinator: H.A.L. van Dijk, EN3S-0045-NL
R. Kübler, Sub-contract to EN3S-0045-NL
P. Achard, Sub-contract to EN3S-0045-NL
S. Furbo, Sub-contract to EN3S-0045-NL
R.H. Marshall, Sub-contract to EN3S-0045-NL
T. Vest Hansen, Sub-contract to EN3S-0045-NL

15.00 - 15.15   Coffee Break

15.15 - 16. 15   CAMUR (Passive Solar Urban Renewal) :

Coordinator: A. Dupagne, EN3S-0048-B
T.M. Maver, Sub-contract to EN3S-0048-B
S. Los, Sub-contract to EN3S-0048-B
M. Raoust, Sub-contract to EN3S-0048-B
P. Geoghegan, Sub-contract to EN3S-0048-B
Y. Michail, Sub-contract to EN3S-0048-B

16.15 - 16.30   PASCAUD (Passive Solar in Computer-Aided Urban Design) :
B.A.M. Welschen, EN3S-0089-NL

16.30 - 16.45   Survey of Building Stock: V. Korsgaard, EN3S-0038-DK

16.45 - 17.30   Control of Passive Solar Systems:

Coordinator: A.H.C. van Paassen, EN3S-0036-NL
P. Bacot, Sub-contract to EN3S-0036-NL
R. Gicquel, Sub-contract to EN3S-0036-NL

Friday, 2 October 1987

09.00 - 09.15   Interzone Airflow: J. Littler, EN3S-0037-UK

09.15 - 10.00   Thermal Comfort in P.S. Buildings:

Coordinator P.O. Fanger, EN3S-0035-DK
I.D. Griffith, Sub-contract to EN3S-0035-DK
P. Depecker, Sub-contract to EN3S-0035-DK

10.00 - 10.15   Coffee Break

10.15 - 10.30   Field-based Research on Thermal Comfort: I.D. Griffith,
EN3S-0090-UK

10.30 - 11.30	Day-lighting:

Coordinator: A. Fanchiotti, EN3S-0047-I
T.M. Maver, EN3S-0047-UK
R. Serra, EN3S-0047-E
G. Willbold-Lohr, EN3S-0047-D
P. Chauvel, EN3S-0047-F
M. Nasi, EN3S-0047-I

11.30 - 12.00	MONITOR: Coordinator: D. Turrent, EN3S-0044-UK
12.00 - 12.15	Monitoring of two houses : Mrs. E. Andreadaki, EN3S-0110-GR
12.15 - 12.45	SOLINFO: Coordinator: J.O. Lewis, EN3S-0087-IRL
12.45 - 14.15	Lunch Break
14.15 - 14.45	ARCHISOL : Coordinator : J.O. Lewis, EN3S-0088-IRL
14.45 - 16.30	Project PASSYS:

Coordinator R. Gicquel, EN3S-0034-F
J.A. Clarke, Sub-contract to EN3S-0034-F
H.J. Reitz, Sub-contract to EN3S-0034-F
J. Uyttenbroeck, EN3S-0029-B
H.A.L. van Dijk, EN3S-0030-NL
L. Bourdeau, EN3S-0031-F
J. Twidell, EN3S-0032-UK
D. Borgese, EN3S-0033-I
V. Korsgaard, EN3S-0085-DK
H.J. Reitz, EN3S-0086-D
J.M. Cejudo, EN3S-0102-E
E. Maldonado, EN3S-A-219-P

16.30	Conclusions

"EUROPEAN SOLAR MICROCLIMATES"

Contract Number   : EN3S-0049-D (B)

Duration   : 36 months 1 April 1986 - 31 March 1989

Total Budget   : DM 2.616,354.--

CEC Contribution   : DM 2.616,354.--

Head of the
Project   : Dr. K. Scharmer, GET mbH

Contractor   : ARGE Bonnenberg + Drescher Ingenieurgesell-
                      schaft mbH/ G E T - Gesellschaft für Entwicklungs-
                      technologie mbH

Address   : ARGE B+D/GET
                      Industriestrasse
                      D - 5173 Aldenhoven

Summary

   The research programme on Eurpean solar microclimates aims for
better prediction of the technically usable solar radiation of a
specific site on the basis of currently available site specific data
i.e. synoptic meteorological data, pollution, orography and
topography.

   At eight characteristic sites precise recordings are done for
global solar radiation, meteorological data and pollution. These sites
are Brussels and Torino (industrial cities situated in plains),
Grenoble and Saarbrücken (industrial towns in narrow valleys), Vendée
in France and the Pentland Hills in Scotland (interaction of coastal
marine and inland climate), Sierra Nevada (strong climatic variations
in mountain valleys)   and Strasbourg (large valley with surrounding
mountains). In the course of the past twelve months the measuring
stations have been put into operation and the test phase has be
accomplished. Despite the bad wheather situation of this year first
results could be achieved and the functionality of the programme
approach and the applied means could be proved.

EUROPEAN SOLAR MICROCLIMATES

1. Introduction to the Research Programme

The costs of solar energy are determined by the magnitude of the initial investment and the amount of technically useful energy harvested during the total lifetime of the installation. In order to reduce investment costs and to maximize the useful energy output it is necessary to predict with high accuracy the site-specific radiation potential, its quality and its frequency in time.

With high effort from the European Community the macro-scale radiation pattern in Europe has been established /1/. The Solar Radiation Atlas for horizontal and inclined surfaces gives statistically reliable values for the average macro- and meso-scale solar potential including daily and seasonal variations.

Nevertheless the local potential may be highly influenced by site-specific effects related to the microclimate as local cloud formation, influence of antropogenetic pollution, specific orographic situation, and may differ considerably from predictions done on the basis of the meso and macro scale radiation levels.

The goal of the European solar microclimatic research is to

- quantify the time and space dependent global solar radiation for typical sites,
- measure the site-specific micro meteorological parameters,
- measure pollutants which may influence the global solar radiation,
- define micro-site specifications with respect to topographic and orographic data,
- correlate the technically usable radiation to effects which may alter the locally available quantity of solar energy and its distribution in time and space in the micro region.

A measuring network was established at eight characteristic sites within six European countries. Table I gives a survey on the participants and their specific tasks in the frame of the overall programme.

From all measuring stations consistent data sets are produced which may be compared to other sites and which are documented in a consistent way in order to allow for comparative studies among the participating groups as well as for interpolation to similar sites for later users.

In each micro region a standard measuring network was installed consisting of

- 9 stations for global radiation measurements (in some cases complemented by sensors for temperature, wind direction and -speed etc.)

- 1 station for global radiation, direct radiation and synoptic meteorological data (temperature, windspeed and -direction, atmospheric pressure, humidity, precipitation). This station gives the meteorological data representative for the micro region.

- 1 station outside the microregion, but as close to it as to be representative for the mesoregion where the microregion belongs to. This station registers the global radiation and the same type of synoptic data as the station inside the microregion.

- For the microregions Torino, Brussels, Grenoble and Saarbrücken pollution data (see Table II) are registered from already operating measuring stations.

Table II gives a survey of the nature and frequency of the recorded data.

Each participating group disposes of an automatic data handling system which provides for a data file which is registered and documented by the coordinator and made available for user applications.

In order to provide for a better layout of solar installations, correlation models and simple-to-handle tools for architects and engineers are developed and tested in the frame of this programme. Figure 1 gives the schematics of the general approach.

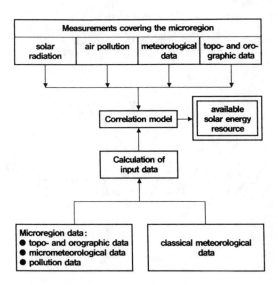

Figure 1: *Establishing the solar energy resources from measured values.*

From detailed measurements of global solar radiation and influencing data resulting from air pollution, from micro-meteorological data and from topo- and orography a correlation model between these data and the technically available solar energy resource is established. This correlation model will allow to define for a given site in a region which is characterized by its site-specific data, the solar energy in quantity, quality and dependence in time by using the appropriate computerized correlation programme and as input data long term meteorological data from "classical" synoptic measurements.

In the frame of the project different existing correlation models are to be tested and evaluated. The necessary input data are to be defined using simple to handle parametrization methods and providing for a tool to establish input data sets from currently available resources.

2. Project Work under Way

2.1 Validation of Instrumentation

In order to quantify microclimatic effects high-precision instrumentation and registration is necessary. All sites are equipped with class I pyranometers from Kipp & Zonen, CM 11, for global solar radiation measurements. From each lot one instrument has been calibrated at the solar center at Carpentras which was used as substandard for local calibration of the other instruments used in the measuring campgne. The calibration of the substandards showed an accuracy with respect to the correction factors given by the supplier of better than ± 0,3 %. Substandardization confirmed these results for the other pyranometers with very few exceptions. Figure 2 shows a typical plot of deviation from a calibrated standard over a days' period.

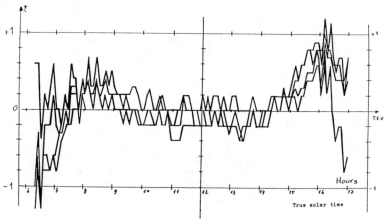

FIGURE 2 : Calibration Scatter Plot for 4 pyranometers CM-11 sampled from 11 CM-11 of the European Communities .

Calibration of pyranometers by comparison with a reference pyranometer based on 6 minutes irradiation measurements .

For practical energy measurements the deviations at small incident angles are of minor importance so that the energy-relevant radiation is registered with an accuracy which is far better than 1 % (see Fig.3)

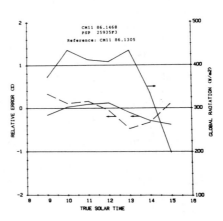

Figure 3: Comparison between one CM 11 and one Eppley PSP and the CM 11 reference pyranometer; Hourly irradiation for the day (temperature of 2 °C)

Positioning, mounting and maintenance of the instruments during the measuring campagne may influence the measurement accuracy. Frequent (at least 6 months interval) checks with an independant substandard will help to eliminate undetected errors. Frequent cleaning and inspection of the individual instruments are performed. Stations operating under harsh conditions (low temperatures, ice and snow) are using instruments equipped with a ventilator to keep the pyrheliometer coup clean at all conditions (Fig. 4.)

Figure 4: Ventilation unit for CM 11 pyranometers (Document Météorologie Nationale)

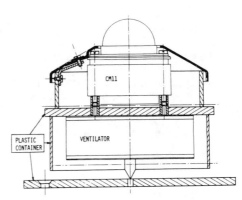

Additionally to the instruments the data processing line has to be calibrated. During the commissioning phase difficulties in time and temperature stability of the registration instrumentation have been encountered and were eliminated. Influence of high frequency fields (emitters) in some cases made it necessary to change locations. In some cases power failures and thunderstorms have let to loss of data and to instrument failure. These difficulties are actually eleminated and most parts of the solar radiation registration instruments are operating satisfactorily.

Synoptic meteorological data are registered according to the rules of the CIMO guide /2/. Calibration checks and yearly recalibration of the whole instrumentation assures the quality of the measured data during the whole measuring period.

Air pollution data are taken over from measuring networks which are operating in parallel to the solar and meteorological network of the project. We are greatful to those organizations who agreed to cooperate with our project and made the data files available.

2.2 Preliminary Results

From measurements performed during the past 3 - 6 months only very preliminary conclusions can be drawn. Most important for the project was to demonstrate the reliability and accuracy of the employed measuring instrumentation, the methods and the software. This result has in all cases been achieved with a high degree of satisfaction. Furtheron the comparison of radiation registered at different measuring points inside the network show that marked variations with space and time occur from one station to the other. These results show the validity of the original approach in selecting the stations as representative for a specific microclimate. Difficulties are encountered with interpretation of cloudiness with respect to the microclimate and the correlation of cloud-affected measurements of substations in the microregion.

Several measurements showed a clear correlation between high level of pollution and variations in Linke turbidity.

2.3 Pollutant Parameter

Measurement of pollutants is comparatively complex and cost-intensive. An approach showed that there is a significant correlation between dust - the most important factor of pollution concerning solar radiation - and SO_2, an easy to measure pollutant (see figure 5). This correlation depends on the type of pollutants, the sources, weather conditions and period of the year. So this correlation actually made for Brussels has to be repeated at other sites during other periods of the year.

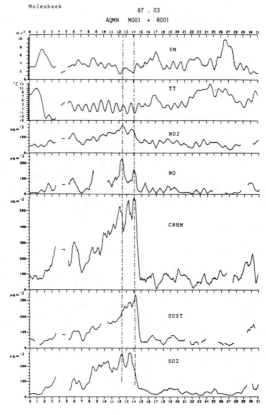

Figure 5:
Air-pollution and
meteorology at Molen-
beek during March 1987
Brussels

3. Site Specifications and approach to Modelling

An approach is under way to characterize each individual site with
a set of parameters in order to achieve a basis for future
extrapolation from the results of this project to other sites. The
models to be used for the correlation of radiation and conventional
meteorological data are being analysed with respect to their
applicability to this programme.

The computer model which is under development will have the
possiblity of using different model structures in order to cover a
large spectrum of possible applications and data sets.

4. Spin Offs

Already now the multidisciplinary character of this project has shown some spin-off effects.

- Meteorology

A discussion is engaged with meteorologists in order to integrate the measurements of this programme into the national grids.

- Instrumentation

Eight different data processing systems are used in the project. A discussion with the manufacturers on weak points in the hard- and software has let to modifications and improvements. In the course of the following year an analysis of the performance will be made stating the minimum requirements for further developmental work.

- Hydrology

Precipitation is measured in 10 minute intervals. This represents an excellent information for dimensioning of sewage systems. Actually these systems are dimensioned using "thump rules" and monthly means of precipitation, which leads to gross errors and normally over-dimensioning of these systems. With these high resolution measurements a much better optimization could be achieved and a tremendous amount of money could be spared.

5. Publication and Information

During last year a press communiqué on the project has been issued by EC DG XII, Brussels.

The project was published at the European Conference on Architecture /3/ and presented on a symposium "Mikroklima in Saarbrücken" on 2nd June 1987 in Saarbrücken, Federal Republic of Germany.

6. Foreseen Activities

The coming winter will prove the functionality and reliability of the instrumentation and the data recording equipment. From September '87 on, all stations should be working continuously and definite data files will be established.

Site-specific data sets (topography and orography) are established and the measurement of the correlation of Linke factor and orographic characteristic will be continued.

Development of models and the parametrization of data to form input-parameter sets will continue and first correlations will take place during summer next year.

Literature

/1/   W. Palz Ed.; Atlas über die Sonneneinstrahlung Europas,
       Kommission der Europäischen Gemeinschaften, Verlag TÜV
       Rheinland, 1984

/2/   WMO N° 8; Guide to Meteorological Instruments and Methods of
       Observation, Fifth Ed. 1983

/3/   K. Scharmer; Solar Microclimates - A concerted research
       programme of the European Communities, European Conference on
       Architecture, Munich 1987

Table I Sites and Type of Microclimate

Location	Type of Microclimate	Contractor	Project Leader
Torino	South-european industrial city with flat environment, high pollution load and frequent inversion weather conditions	Fiat Engineering Spa.	Mr. Giacomelli
Brussels	industrial city with low, hilly environment and influenced by the north-sea climate. High rate of pollution	Institut Royal Météorologique de Belgique	Mr. Joukoff
Grenoble	industrial city located in a Y-shaped alpine valley, influenced by mountain borne valleywinds, frequent thermal inversion and high pollution	Université Scientifique et Médicale de Grenoble	Mr. Grenier
Saarbrücken	industrial city located in a valley basin cut into the surrounding plateau. The mild microclimate of this valley is strongly influenced by the agglomeration and the local heavy industry	Landeshauptstadt Saarbrücken, Amt für Energie und Umwelt	Mr. Luther
Upper Rhine Valley (Strasbourg)	The large flat basin of the Upper Rhine Valley protected through mountain chains shows a typical microclimate with marked variations in sunhours	ADURE Strasbourg	Mr. Hug
Sierra Nevada (Granada)	The steep-raising mountains show in a very limited space climatic variations from subtropical to alpine character	Department of Fundamental Physics Faculty of Sciences, University of Granada	Mr. Jimenez
Vendeé (France)	Sea-side microclimate influenced by the atlantic and sea-side cloudiness variations	Direction de la Météorologie Nationale, Paris	Mme Villien
Pentland Hills	South-west of Edinburgh, a microclimate strongly influenced by the interaction of a coastal marine climate and an inland climate affected by a range of hills	Scottish Centre of Agricultural Engineering Midlothian, Penicuik	Mr. Spencer
Zürich (special measurements		Frei, Schneider + Guha AG	Mr. Valko
Athens (computer model)		Lamda Technical Ltd., Athens	Mr. Lalas
Aldenhoven (coordinator)		ARGE B+D/GET Bonnenberg + Drescher Ingenieurgesellschaft mbH/ GET - Gesellschaft für Entwicklungstechnologie mbH	Mr. Scharmer

Table II Data Records

measured quantity	kind of measured value	recording	dimension
global radiation (CM11)	A	10 min	W/m²
air temperature	I	10 min	1/10 °C
relative humidity above ground (roof) level	I	10 min	%
(roof) level			
atmospheric pressure	I	10 min	10 Pa
wind speed	A	10 min	1/10 m/s
wind direction	V	10 min	10 °C
precipitation	S	10 min	1/10 mm
cloudiness	I	180 min	1/8
direct radiation (CM 1)	A	10 min	W/m²
atmospheric pollution	S	30 min	μ/m
SO , NO, NO , CO , OH, O , TSP			

V   = vectorial resultant
I   = instantaneous value
A   = average value
S   = integral value
TSP = total suspended particles

DATA ON SOLAR RADIATION AT EARTH
FROM SATELLITE MEASUREMENTS
(PROJECT SUNSAT)

Contract Number   : EN3S-0112-D (B)

Duration   : 30 months   1.7.1987-31.12.1989

Total Budget   : DM 1.011.500,-

Head of Project   : Prof. Dr. E. Raschke, Institute of
                     Geophysics and Meteorology

Contractor   : University of Cologne, F.R.G.

Address   : Institut für Geophysik und Meteorologie
                     der Universität zu Köln
                     Kerpener Straße 13

                     5000 Köln 41
                     F.R.G.

Summary:

    The global radiation over the entire African continent
will be monitored with satellite data. Final results
show errors less than 10% with a spatial resolution of
30-50 km if averaged over periods of 10-15 days.
During this project data of the satellite METEOSAT 2
from a period of at least 2 complete years (1985 and
1986) will be evaluated. Careful validations of the
results will be made intercomparing them with
simultaneous and co-located ground-based measurements.

1. Introduction: The need for data

Statistical data on available solar radiative energy over each area of the earth are required for a large variety of purposes. These are interrelated and may be grouped into two categories:

utilizations: renewable energy sources (solar energy)
                    agriculture   } "biomass production"
                    forestry   } and others
                    architecture
                    holiday industry

research: properties of climate on earth
                    studies of climate processes
                    weather system forecasts

These applications require different kinds of data, such as
- radiation onto horizontal surfaces
- radiation onto inclined surfaces
- spectral components of diffuse and direct radiation.

Furthermore, these should be available with different resolutions in time, e.g. as
- daily sums and statistics
- hourly sums (or higher) and statistics

Further easy access should be available for the user. The operations of more sophisticated solar power systems and also other needs call now even for the availability of predictive models covering the "next sequence" of hours and days.

2. Data sources:

Direct measurements of these required quantities are only sporadically possible, since they require intensive manned controls of instruments. Thus the earth's continents are covered only with sparse networks where largest gaps occur within all countries not belonging to the industrially "developed world". But these are of major interest for the project.

Therefore intensive research began - also supported within earlier programmes of the CEC - to fill these various gaps between different data sets
- deriving data from spatially and temporally dense satellite measurements, and
- deriving data from others via computational procedures.

The results of such methodologies are rather accurate and may in many cases already be even accurate enough to satisfy the needs of the data user. However, in several applications - e.g. the planning of solar power plants at specific locations - they should be complemented by series of additional direct measurements.

These latter must also provide the temporal and spatial details which are required to plan and operate efficiently such a power plant. Some data banks already exist, mostly at national weather services but also at several relevant research and environmental establishments.

3. Solar radiation from satellites:

Since clouds reflect most dominantly the incident solar radiation back to space and absorb only a relative small fraction of it, not exceeding 10-15 percent in most cases, there exists a narrow but inverse regression between the amount of solar radiation reaching ground and the cloud cover and optical thickness of clouds. The latter properties of cloud fields also modify the spectral properties of radiation, and its partitions into direct and diffuse components. These are determined by often rapidly varying concentrations and optical properties of atmospheric aerosols.

In Europe - and primarily supported by a related project (Solar Energy R+D in the European Community, Series F) of the CEC several groups developed and tested methods to calculate the amount of solar radiation (= global radiation) reaching horizontal surfaces at ground from the operationally available imaging data of meteorological satellites. These are digital imaging measurements made within the solar spectral region to identify primarily cloud fields within atmosphere.

Of particular value for these purposes are measurements of geostationary satellites which are made with spatial resolutions ranging from 2-5 km per pixel and with repetition rates ranging from 30 minutes to 2 hours. These high repetition rates enable valuable considerations of the temporal variability.

For Europe and Africa the measurements of the European geostationary satellites METEOSAT 1 (1977-1979) and 2 (1981-now) and of its successor METEOSAT 2 provide this useful data base. However, there is no continuous archive of data with original high spatial resolution. But the ESOC in Darmstadt keeps data probes and a continuous archive of sampled data (Δx, Δg 30 km, Δt = 3h). It could be shown, that monthly averages of daily sums of total solar radiation could be derived from them with accuracies ranging from 5 to 8 percent.

4. The project SUNSAT:

Within the programme on renewable energy sources the CEC supports the project SUNSAT, whose preliminary objective are

-   to determine with a resolution of about 30-50 km and 3 h
    the statistical   properties   of   monthly   total   solar
    radiation fields   at ground over the entire continent of
    Africa for a 2 year period
-   to determine   some quantities for selected months of the
    same period   over the Sahel zone but with higher spatial
    (5 to 10 km) resolution, and
-   to validate carefully these results with respect to this
    accuracy making use of simultaneous measurements at many
    ground stations within all African countries.

    This project   began on   1 July 1987 and will end on 31
December 1989. By then the following data will be available
for further use either on tape or in tabular form:

-   monthly averages   of   daily   sums   of   total global
    radiation,
-   the same of the diffuse solar radiation,
-   monthly averages of global radiations at 3 hourly
    intervals (6, 9, 12, 15, 18 hours universal time),
-   the same of the diffuse component

for the   entire continent of Africa and also Europe as seen
from Meteosat   and also   for the Sahel zone in more spatial
detail. In the latter case only data of January and July of
each year will be analyzed.

    These satellite   derived data   will be complemented by
statistical   analyses   where   they   are   compared   with
simultaneous ground-based measurements. The   latter, since
stemming from   different sources   and taken   with   possibly
different stages of care, undergo a very careful validation
procedure.

    Many data   samples will also be published in pictorial
form in an atlas.

    The following groups participate in this project:
prime contractor: University of Cologne, Germany
subcontractors: University of Wageningen, The Netherlands
                  CTAMN Armines, Sophia Antipolis, France
                  SODETEG, Sophia Antipolis, France
                  Institute of Renewable Energy Sources,
                  Madrid, Spain.

5. Final remarks:

    No results   are available   yet from this project since
it has   started its activities just very recently. During a
workshop in   middle of   October 1987   preliminary   data   as
obtained from data sets will be discussed.

6. Some selected references:

Grüter, W.,   H. Guillard,   W. Möser, J. Monget, W. Palz, E.
        Raschke, R.   Reichardt, P. Schwarzmann, L. Wald,
        1986:   Solar   Radiation Data   From Satellite
        Images.   Solar   Energy R+D   in   the European
        Community, Series F,   Vol.4,   D.   Reidel Publ.
        Comp., 100 pp.

Möser, W.,   E. Raschke, 1984: Incident solar radiation over
        Europe estimated from METEOSAT data. J. Climate
        and Appl. Meteor., 23, 225-234.

EUFRAT:

UTILIZABILITY AND CUMULATIVE FREQUENCY CURVES OF SOLAR IRRADIANCE

Contract Number : EN3S-0111-F

Duration : 24 months   (1 July 1987 - 30 June 1989)

Total Budget : 179 000 ECU

CEC Contribution : 118 000 ECU

Head of Project : Dr. Jérome ADNOT, Centre d'Energétique
Ecole Nationale Supérieure des Mines de Paris

Co-ordinator : Dr. Bernard BOURGES, Consultant
6, rue de l'Armor - 35760 ST GREGOIRE (France)

Contractor : ARMINES - Centre d'Energétique

Address : 60 Bd St-Michel 75272 PARIS CEDEX 06 (France)

Summary

The aim of this concerted Research Action is to prepare, - in
addition to already available data sources such as the Solar
Radiation European Atlas -, climatological data required as an input
for solar system design tools. Five laboratories are involved in this
programme.
The basic objective will deal with the frequency distribution of
solar radiation and the derived concept of Utilizability. Cumulative
Frequency Curves of solar irradiance and Utilizability data sets will
be prepared for a sample of reference locations, from long-term
weather data files. These reference CFC's and Utilizability data sets
will be used in order to establish simple reconstitution formulas.
Some additional related topics will also be investigated: incidence
angle modifier, temperature data, time series aspects, etc. Final
results, - reference data sets and computation methods-, will be
published both as an Atlas and as a software package.

1 - INTRODUCTION

Using solar radiation data in the field of solar energy applications is not only a problem as regards the reconstitution of useful information from what is currently available but also as regards suitable presentation. Generally speaking, any computation methodology requires given weather data processing: from the simple mean monthly irradiations used in a number of very simplified methods, to the "Test Reference Years", the typical yearly weather data files used as an input for detailed solar system simulation programmes (see table I).

Monthly frequency distribution of solar irradiance is now another common intermediate meteorological input for a number of simplified design methods for solar systems such as

- Active thermal systems (Space Heating, Domestic Hot Water, Industrial Process Heating etc.): methods ESM1, Fi-f-charts, SEU etc. (1-2-3)
- Passive systems (Direct Gains, Trombe Wall): Unutilizability method (2)
- Photovoltaic systems: PV-charts method (2)

One of the Frequency Distribution applications is the computation (by a simple numerical integration) of the so-called "Utilizability", i.e. the fraction of incident solar energy actually available above a given irradiance threshold for a certain period. Frequency Distribution, of course, includes information on the average value, but keeping a simple form, it also provides much more complete information and permits much more accurate system performance computation. Moreover, it remains a general statistical presentation (i.e. it is not too specific to a given application), while maintaining a climatic meaning.

Table I: Solar system computation methods and required weather data

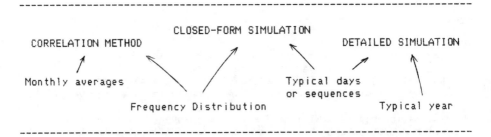

Work has already been done on Frequency Distribution and Utilizability, but, as we shall see, some elements are still needed as regards

- the generalization of former results in terms of climatic area (with some emphasis on Southern Europe, rather neglected until now), direct and global radiation, long-term periods, etc.

- the collecting and processing of additional weather data required for
system design (temperature, cooling and heating degree-days etc.)
- complementarity with available information, such as solar atlasses.
- suitable presentation of the results for use by a wide range of
professionals (Appendix to the European Solar Radiation Atlas, software
packages etc.).

The present Concerted Research Action will deal with these topics.

2 - PRESENT STATE OF KNOWLEDGE IN THE FIELD

As far as Utilizability is concerned, reconstitution correlations are
available, but they are of use only in the computation of Utilizability
itself and not of overall distribution (4-5). Furthermore, they are
derived from North American radiation data and their application to
European climates often gives poor results (6). Generally speaking, they
are based on the implicit assumption that diffuse sky radiation is
isotropic. Very simple correlations nevertheless seem promising (7).

The Frequency Distribution of solar irradiance has been studied (within
the framework of the former R/D European programme) as Cumulative
Frequency Curves (CFC's) of Solar Irradiance for Northern Europe (8).
CFC's have been computed for 12 European locations (from Test Reference
Years, - TRY -, weather data files) (9) and 12 French locations (using
long-term data files for 5 to 10 years) (10). These CFC's have been
computed using an accurate algorithm by Pérez (11) for the calculation of
mean hourly global irradiance on tilted planes from data measured on a
horizontal plane (Errors within 3% for the mean monthly values).

Semi-empirical formulae have been derived from a statistical analysis of
European (TRY) CFC's (12): the Frequency Distribution of solar irradiance
may be computed from three parameters (mean monthly irradiation on the
given surface, maximal irradiance, and time between sunrise and sunset).

As sources of data for users, there are atlasses, such as the European
Solar Radiation Atlas (13) or some national or regional data books. They
include mean monthly irradiations on various surfaces, as well as clear
day data or some kind of presentation of the distribution of daily
irradiation.

3 - UTILIZABILITY AND CUMULATIVE FREQUENCY CURVES OF SOLAR IRRADIANCE

Cumulative Frequency Curves of global solar irradiance are defined as
curves giving the time n_H (mean daily number of hours, during the
month considered), during which the solar irradiance on a given plane has
exceeded a value I_c (threshold irradiance or critical intensity).
Examples are given in figure 1: the time, n_H, is read on the
horizontal axis (Unit: hours/day); the critical intensity, I_c, is read
on the vertical axis (Unit: W/m2).
From this example, it can be seen that intensities greater than 300W/m2 on
a tilted plane (45°) were observed in Carpentras for a period of 9.2 hours
/day in July and only for 0.9 h/d at Kew during the month of December.

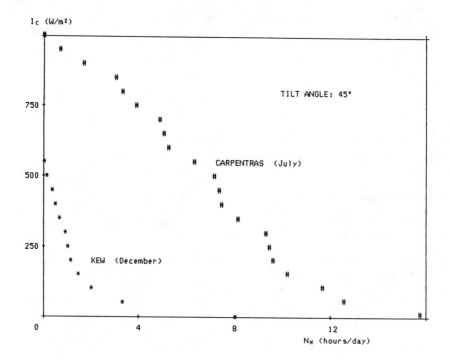

Fig. 1: Examples of Cumulative Frequency Curves of Solar irradiance

Some properties of CFC's should be noted.
The CFC intercept with horizontal axis is the mean number of hours between
sunrise and sunset, d_J.
The intercept with the vertical axis represents the maximal solar
irradiance, I_{MAX}, which can be achieved during the month on the plane
considered. For south-oriented planes, I_{MAX} is generally observed at
solar noon on clear days.
The area limited by the CFC, the vertical axis and the horizontal line I =
I_c represents the solar energy available above the critical intensity
I_c, $H_{AV}(I_c)$. It is given by the simple integral

$$H_{AV}(I_c) = \int_{I_c}^{I_{MAX}} n_H \cdot dI \qquad (\text{Unit: Wh/m2.day})$$

The overall area limited by the CFC and both horizontal and vertical axis
is simply the mean daily sum of global solar irradiation on the plane
considered

$$H_{AV}(0) = H$$

The ratio $\emptyset = H_{AV}(I_c) / H$ is the "Utilizability" defined by

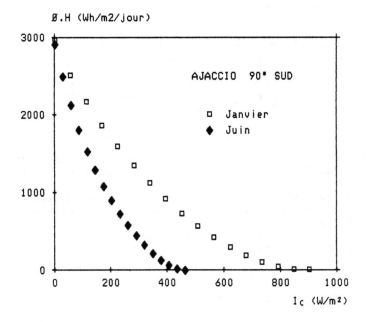

Fig. 2a: Available energy above threshold irradiance

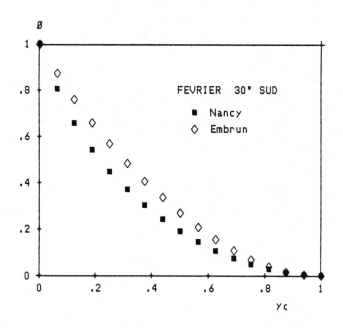

Fig. 2b: Utilizability vs reduced threshold irradiance I_C/I_{MAX}

Liu and Jordan (14) and Klein (4). The Utilizability is the fraction of incident monthly solar irradiation, actually available above a given critical intensity I_C (Fig. 2).
In the case of a flat-plate solar collector (Optical efficiency: n_0; Loss coefficient: U_L) operating at Temperature T_B, with ambient temperature T_A, the critical intensity is defined by

$$I_C = U_L / n_0 . (T_B - T_A)$$

The Cumulative Frequency Curve of a given month also enables to define a "Typical Meteorological Day" which reproduces both the frequency distribution and a realistic time profile (Fig. 3).
Cumulative Frequency Curves of solar irradiance are obtained by a sorting of hourly files of solar global irradiance. For horizontal planes these data are measured by meteorological services in various locations. But, as far as inclined surfaces are concerned, data are rare and have to be computed from global and diffuse (or direct) radiation on horizontal plane.

4 - AIM OF THE PROPOSED R/D WORK

The main objectives of the study are based on the following considerations:

   - Utilizability and frequency distribution of solar irradiance are a fundamental factor for solar system design methods. They must be made available in a form suitable for a wide-ranging population of users.
   - CFC's and Utilizability reference data sets have to be made available for a sample of representative European locations (Northern and Southern as well) from reliable long-term solar radiation measurements.
   - Not only flat-plate collectors are concerned: concentrating or semi-concentrating devices such as CPC's should also be included.
   - General and simple results are expected, in terms of CFC's and Utilizability correlations for any European climate.
   - All the climatic information needed by designers has to be concentrated; the final publication should include additional items, such as temperature or degree-days data, consistent with the solar radiation data.

Thus, the concerted action will deal with the frequency distribution of solar radiation, utilizability and some related topics, with the aim of producing a reference document, used as an appendix to the European Solar Radiation Atlas.

In order to achieve these general objectives, the following tasks are to be completed:

   - CFC's and Utilizability data set.
   CFC's and Utilizability for various types of surfaces and radiation (global, direct) will be computed from available hourly solar irradiance data files.

- CFC's and Utilizability correlations.
From the former data sets, reconstitution correlations of CFC's and
Utilizability will be developed, in connection with already existing
data sources such as the European Atlas. Specific correlations will be
proposed for monthly irradiations and maximal irradiances.

- Additional tasks.
Some other related items will be investigated: incidence angle
modifier; collecting and processing temperature data; introduction of
time series aspects; frequency distribution of solar irradiance for
hourly periods.

- Presentation and publication of results
All the results (methods and data tables) will be compiled in one
volume. Suitable software programmes will also be developed.

From the development of reliable summarized solar radiation data sets
(simultaneously with the improvement of simplified design methods within
the framework of other projects), the present study should result in the
improved design of solar systems by engineers and manufacturers. It will
produce a useful complement to the actual European Solar Radiation Atlas.
Its applicability could likely be extended to other areas with a large
solar potential (Africa, Middle East etc.).

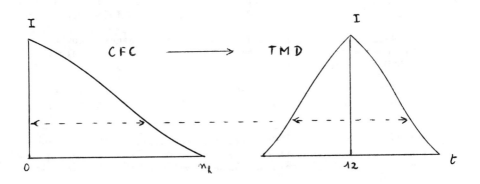

Fig. 3:   From a Cumulative Frequency Curve to a Typical Meteorological Day

REFERENCES

/1/ J.ADNOT, B.BOURGES, B.PEUPORTIER, W.DUTRE and T.C.STEEMERS. Design and validation of simplified methods for sizing active solar space heating systems. First EC conference on Solar Heating, Amsterdam (1984).

/2/ S.A.KLEIN and W.A.BECKMAN. Review of Solar Radiation Utilizability. ASME Journal of Solar Energy Eng., Vol. 106, pp.393-402 (1984)

/3/ J.P.KENNA. A parametric study of open (closed) loop solar heating systems. SOLAR ENERGY, Vol. 32, n°6, pp.687-723 (1984).

/4/ S.A.KLEIN. SOLAR ENERGY, Vol.21, n°5, pp.393-402, (1978).

/5/ M.COLLARES-PEREIRA and A.RABL. Simple procedure for predicting long-term average performance of nonconcentrating and of concentrating solar collectors. SOLAR ENERGY, Vol. 23, pp.223-253 (1979).

/6/ M.J.CARVALHO and B.BOURGES. Application of Utilizability Computation Methods to Europe and Africa. Congrès INTERSOL 85, Montréal (1985)

/7/ M.J.CARVALHO, M.COLLARES-PEREIRA, J.M.GORDON and Y.ZARMI. Yearly and monthly Utilizability functions for principal solar collector types. SOLAR WORLD CONGRESS, Perth. Vol.4, pp.2219-2224, (1983). (Published by Pergamon)

/8/ B.BOURGES. Statistical Distribution of Solar Radiation: A European Data Set of Cumulative Frequency Curves of Solar Irradiance on Tilted Planes. Int. J. Solar Energy, Vol. 3, pp.241-253 (1985).

/9/ B.BOURGES and F.LASNIER. Statistical distribution of solar radiation: Cumulative Frequency Curves. CEC Report ESF 008-F, ARMINES, Paris (1983).

/10/ B.BOURGES. Atlas de l'Energie Solaire pour la France: Courbes de Fréquences Cumulées de l'Eclairement Solaire et Energie Disponible. Rapport AFME 3.350.3168 (1985).

/11/ R.PEREZ et al. A new simplified version of the Perez diffuse irradiance model for tilted surfaces. SOLAR ENERGY, Vol. 39, n°3 , pp.221-232 (1987).

/12/ B.BOURGES and M.GRIECH. Utilizability Computation from Frequency Distribution of Solar Irradiance: Preparation of a European Data Set and Development of Empirical Formulae. Congrès INTERSOL 85, Montréal (1985).

/13/ W.PALZ et al. Solar Radiation European Atlas (2 volumes). Verlag TUV Rheinland, Cologne (1984).

/14/ B.Y.H.LIU and R.C.JORDAN. SOLAR ENERGY Vol.7, n°2, pp. 53-74, (1963).

/15/ D.FEUERMANN, J.M.GORDON and Y.ZARMI. A typical meteorological day approach for predicting the long-term performance of solar energy systems. SOLAR ENERGY, Vol.35, n°1, pp.63-69 (1985).

Table II: Summary of the EUFRAT research work planned

Phase 1: Computation of CFC's and Utilizability data sets

- Development of a FORTRAN programme to be used by all the participants.
- Further validation of computation algorithms for global radiation on tilted planes.
- Collection and control of radiation data files (hourly values) from various European countries.
- Computation of Utilizability and Cumulative Frequency Curves of global solar radiation on planes of different orientation and tilting and of direct radiation for different types of concentrators.
- Gathering of data sets on one single tape.

Phase 2: CFC's and Utilizability correlations

The former data set will be used for developing computation correlations of

- Utilizability (monthly and yearly)
- Frequency distribution (CFC's)
- Mean monthly and yearly irradiation on collector surfaces
- Maximal irradiance

The accuracy and validity range of these correlations will be characterized.

Phase 3: Additional tasks

- Incidence angle modifier. Introduction into Utilizability computations.
- Additional data collection and processing: Temperature, degree-days etc..
- Time series.
- Hourly frequency distribution of irradiance.

Phase 4: Publication of results.

Basic results will be presented in a easy-to-use publication which could be an useful Appendix to the European Solar Radiation Atlas. The link with the use of Solar Irradiance Frequency Distribution, namely in simplified solar system design methods, will be established and developed. This report will include complete data tables and, if required for some of the parameters, maps. It will be prepared in the form of "camera-ready copies".
Conclusions of the work resulting in calculation methods could also be reported in the form of software packages for computers.

Table III: Timing of the research proposed

		TIME	(MONTHS)		
PHASE	0	6	12	18	24
	.	.	.	.	.

1
(CFC's and Utiliz. data set)   <-------------->

2
(CFC's and Utiliz. correl.)   <-------------->

3
(Additionnal tasks)   <--------------------->

4
(Final report + Atlas)   <--------->

Starting Date: 01 July 1987   Duration: 24 months

Table IV: EUFRAT participants

CONTRACTOR: ARMINES — 60 Bd St Michel   75272 PARIS CEDEX 06   (FRANCE)
               Centre d'Energétique

CO-ORDINATOR: B.BOURGES — Consultant
               6, rue de l'Armor   35760 ST-GREGOIRE   (FRANCE)

SUB-CONTRACTORS:

        LNETI. (Laboratorio Nacional de Engenharia e Tecnologia
        Industrial)   Dpto de Energias Renovaveis
        Estrada do Paço do Lumiar, 22   1600 LISBOA   (PORTUGAL)

        Escuela Superior de Ingenieros Industriales de Sevilla
        Av. Reina Mercedes   —   41012 SEVILLA   (SPAIN)

        Facultad de Ciencias Fisicas. Grupo de Energia Solar
        Ciudad Universitaria.   28040 MADRID   (SPAIN)

        National Observatory of Athens.   Meteorological Institute
        P.O.BOX 20048   —   11810 ATHENS   (GREECE)

O P S Y S

_____

CONCERTED ACTION FOR SOLAR SYSTEM MODEL DEVELOPMENT AND VALIDATION

_____

Contract Number: EN3S-0046-B
Duration   : 24 months, 1 June 1986 – 31 May 1988
Total budget : 26 MBF
Contractor   : Katholieke Universiteit te Leuven
Coordinator   : W.L. Dutré
                Mechanical Engineering Institute, K.U.Leuven
                Celestijnenlaan, 300 A,
                B-3030 Heverlee, Belgium

Subcontractors: - Ecole des Mines de Paris (France)
                - Technisch Physische Dienst (Delft, The Netherlands)
                - Thermal Insulation Laboratory (Lyngby, Denmark)
                - E.E.S. (The Netherlands)

Participants : - W.L.Dutré, A.Debosscher, K.Lambrechts (K.U.Leuven)
                - J.Adnot, B.Peuportier, B.Bourges (Ecole des Mines)
                - A.De Geus, P.Bergmijer (T.P.D.)
                - S.Ostergaard-Jensen (D.T.H.)
                - C.Den Ouden (E.E.S.)

Summary

      The main objectives of the Opsys program include the validation of
EMGP2-simulation models for various SPTF-SS2 systems, the development of
user friendly software packages for system simulation and simplified
correlations for system design. The validation has been completed for
the SPTF-SS2-Belgium system and is proceeding for the Danish and Dutch
SPTF-SS2-configurations, which include different types of storage tanks.
From the results obtained until now, no systematic model errors have
been detected and it can be stated that the simulations based on
reliable sets of system characteristics, agree with experimental
results. The observed discrepancies are within the calculated uncertain
margins. With respect to the development of simulation programs based
on interactive procedures, a complete modular simulation program package
is available consisting of two programs and various data files and help
files. The system specific simulation program has been extended with
various additional features and system configurations. The final
version of the simplified correlation program is being developed and a
new approach for a simplified method for medium sized solar water
heaters is investigated.

1. INTRODUCTION

The concerted action OPSYS started in June 1986 and aims at the development of simulation models and simplified correlations for the analysis and performance calculations of active thermal solar systems and the validation of these computational tools. Such developments were already initiated by the European Modelling Group,a concerted action of the C.E.C. in the period 1981-1983, of which the final products were the first main frame computer version of the modular simulation program EMGP2 for transient solar system simulations, a first version of the simplified correlation method ESM1 for space heating systems and a comprehensive set of validation calculations for the common solar system of all SPTF-installations, known as SPTF-SS1. The results of this first validation campaign were satisfactory but were limited to solar systems with non-selective collectors and unstratified storage tanks for space heating only. It was therefore decided to continue the model validation activity for other types of systems and components, based on the experimental data of some of the SPTF-SS2's, obtained in course of the Solar Pilot Test Facility project, another concerted action of the C.E.C. during 1979-1983. It was also decided that user friendly computational design tools for active thermal solar systems, based on the EMGP2-component models should be developed.

The objectives of OPSYS therefore include three types of subtasks:
- validation of EMGP2-component models and system models, for the selected SPTF-SS2-systems.
- The development of solar system simulation software as user friendly design tools which can be used on personal computers.
- to extend the simplified correlation method ESM1 to combined systems for space heating and hot water production and to develop a separate simplified correlation method for solar water heaters.

These general objectives, the corresponding subtasks of the Opsys concerted action and the expected final products are described in more detail in [1], a paper presented at the E.C.-contractor's meeting held in Brussels in november 1986. The present paper will therefore concentrate on the validation results already obtained and the software developments which have been achieved.

2. VALIDATION OF EMGP2-MODELS.

The validation subtask of OPSYS is devoted to the validation of EMGP2-models for the following SPTF-systems:
- SPTF-SS2-Belgium: a combined solar system for space heating and hot water production, as represented in figure 1. The system includes a heat pump and a secondary storage tank as a buffer between the solar system-heat pump combination and the space heating demand. Domestic hot water preheating is achieved by means of a preheating tank submerged in this secondary storage tank.
- SPTF-SS2-Denmark: a solar system which has successively been equiped with three different types of storage tanks, as represented in figure 2. Configuration SS2.1 includes a preheating tank for domestic hot water, submerged in the solar storage tank (SS2.1). Configuration SS2.2 contains a horizontal storage tank with a submerged heat exchanger coil and in configuration SS2.4, a jacketted storage tank is used.

- SPTF-SS2-The Netherlands: a combined system for space heating and tapwater heating, equiped with evacuated high performance collector tubes. Two different configurations of this system have been investigated: a configuration with a stratified liquid storage tank with submerged heat exchanger and a second configuration with a latent heat storage. The emphasis of this validation task is mainly put on the latent heat storage system.

2.a. Validation procedure.

The validation calculations are mainly concerned with the energy transfers at different levels of the considered systems: collector array energy output, storage tank energy input, storage tank energy output to space heating and to the tapwater load. Since the European Modelling Group experienced that validations for complete systems only, do not yield sufficiently detailed information to evaluate the validity of main system component models and configurations by which they are represented in the simulation calculation, the validation procedure applied by the Opsys-participants, mainly emphasizes subsystem and component validations. To this extend, the considered systems are subdivided in a number of subsystems, such as the collector subsystem and the storage subsystem, accounting for the location of the measurement sensors from which the experimental values of the energy quantities are calculated.

According to the methodology previously used by the European Modelling Group, the uncertainty margins of the experimental and the theoretically calculated energy quantities as well as the uncertainty margin of the observed discrepancies, are calculated for all subsystems and data sequences considered. This calculation is based on a linear error propagation assumption and calculations of the sensitivity of each considered energy quantity to the meteorological input data, measured temperatures and flow rates used as input data and to the various system parameters. The method is presented in [2]. For the experimental energy values, the uncertainty margins result from unknown systematic measurement errors, while the uncertainty to be associated with the results of the simulations, are the combined effect of the uncertainty margins of the descriptive system parameters and the measured quantities used as input data for the simulation. In the calculations of the uncertainty margin of the observed discrepancies, allowance is made for the fact that the experimental and theoretical values are not fully independent: some of the measured data used to evaluate the experimental energy flows, are also to be used as input data for the simulations. It follows that the corresponding error margins are correlated. The uncertainty margins allow to verify whether the observed discrepancies are within the normally expected range as a result of uncertainties of all input data and parameters or indicate significant model errors.

Since several system parameters can not be measured or calculated with a sufficient degree of accuracy, the validation procedure includes a parameter calibration step based on a representitive experimental data sequence, which is thereby considered as a transient measurement of some system parameters and excluded from the set of data sequences used for actual validation purposes.

2.b. Present state of the validation subtask.

For SPTF-SS2-Belgium all validation calculations and uncertainty analyses have been completed. This set of validation calculations includes 28 experimental data sequences, totalizing 268 days of measurements with a time step of five minutes. For the storage subsystem and the collector+storage subsystem, some results have been included in [1]. Final results of the simulation calculations for the entire system represented in figure 1, are shown in de following figures, in which the calculated energy values versus the experimental values are plotted:

figure 4:   the collector array energy output
figure 5:   the main storage tank solar energy input
figure 6:   the total energy output of the main storage tank
figure 7:   the heat pump evaporator energy input from the storage tank
figure 8:   the secondary storage tank energy input from the heat pump condensor loop.
figure 9:   the secondary storage tank energy output to space heating
figure 10: the submerged tapwater preheating tank energy output.

Some corresponding uncertainty margins of the observed discrepancies are shown in the figures 11, 12 and 13. As compared to the subsystem validation results given in [1], it can be seen that the discrepancies of the complete system simulations are significantly larger. This is due to the cumulated effect of systematic errors on all the system parameters used as input data for the simulations. These effects also result in correspondingly larger uncertainty margins for the energy quantities and their discrepancy from the measured values. Most of the observed discrepancies remain however within the one standard deviation interval and all values are within the three standard deviations interval. Although the population of validated sequences is not large enough for statistical laws to be fully applicable, these results indicate that systematic model errors are small as compared to the effect of parameter uncertainties and systematic measurement errors.

For the SPTF-SS2-Denmark, validation calculations have been performed for system configuration SS2.1 shown in figure 2. The main problem encountered is related to the tapwater draw-off from the submerged preheating tank. The system clearly suffers from strong mixing effects and a partial short-circuiting between the cold water inlet en the preheating tank outlet. Such phenomena strongly depend on the pressure of the cold water feed, the draw-off flow rate, the position of the inlet pipe in the preheating tank and the internal flow pattern. Unless much more sophisticated models would be used, these effects can be described only empirically. Applicable EMGP2-configurations which allow to account for such effect by means of empirical factors, are being investigated. For the collector energy output, the storage energy input and the storage energy output to space heating, the results obtained thusfar are satisfactory, as be seen in table I hereafter. The results of the corresponding uncertainty analyses are represented in the figures 14, 15 and 16, in which the full lines only connect the plus and minus one standard deviation values for the different data sequences indicated along the abscissa of these figures. Nearly all data points are within the one standard deviation interval, except for the data sequence with small daily solar energy input. Such sequences are not representative for normal system operation and have no significant influence on long term system performance. The corresponding uncertainty margins are percentagewise also much larger, due to operation of the system at small temperature differences. These results are briefly described in [4].

For configuration SS2.2 of SPTF-SS2-Denmark, the main difficulty encountered is the construction of a suitable EMGP2-configuration to represent the horizontal storage tank with an acceptable degree of accuracy, without having to represent the internal flow pattern. The results obtained from the validation runs performed thusfar, are given in table II. Validation results for the SS2.4 configuration are not yet available.

For the SPTF-SS2-The Netherlands, starting from the initial set of system parameters established during the start-up of the installation, a collector parameter modification has been shown to be necessary. This modification of the collector optical efficiency and the back side heat loss coefficient, is probably due to a short term ageing effect of the high performance collectors being used, which stabilizis after some months of operation. To determine the valid set of parameters, one of the available data sequences has been used. Available data sequences for validation include 239 days of operation, with hourly averaged measurements. Validation results are not yet available.

Data seq.	Coll.Energy Output			Storage Energy Input			Storage Energy Outp.		
	meas.	calc.	% err	meas.	calc.	% err	meas.	calc.	% err
2004	1233	1251	1.5	1184	1216	2.7	878	897	2.2
2005	1465	1453	-0.8	1386	1403	1.2	791	809	2.3
2006	1402	1496	6.7	1434	1480	3.2	1096	1143	4.3
2008	2111	2153	2.0	1991	2031	2.0	1268	1243	-2.0
2009	1692	1769	4.6	1617	1712	5.9	698	702	0.6
2010	760	879	16.0	729	860	18.0	442	458	3.6
2011	1033	1195	16.0	1037	1176	13.0	643	699	8.7
2012	264	343	30.0	262	356	36.0	226	242	7.1

Table I: SPTF-SS2.1-Denmark: Measured and calculated energy flows (expressed in Megajoules) and percentage discrepancy.

Data seq.	Coll.Energy Output			Storage Outp.to SpH.			Storage Outp to DHW.		
	meas.	calc.	% err	meas.	calc.	% err	meas.	calc.	% err
D2027	1725	1726	0.1	1446	1318	-8.9	146	140	-4.1
D2029	1003	1064	6.1	793	751	-5.3	261	232	-11.0
D2031	1518	1673	10.0	948	874	-7.8	290	294	1.4
D2032	1401	1546	10.0	1025	938	-8.5	162	174	7.4
D2033	1057	1139	7.8	897	788	-12.0	168	163	3.0
D2034	1315	1501	14.0	522	502	-3.8	294	293	-0.3
D2036	757	785	3.7	0	0	0.0	438	411	-6.2
D2037	740	775	4.7	507	418	-18.0	169	169	0.0
D2038	1158	1226	5.9	953	848	-11.0	152	143	-5.9
D2039	451	479	6.2	346	320	-7.5	87	89	2.3
D2041	229	362	58.0	50	65	30.0	148	160	8.1

Table II.: SPTF-SS2.2-Denmark: Measured and calculated energy values (expressed in Megajoules) and the percentage discrepancy.

3. DEVELOPMENT OF SIMULATION PROGRAMS.

As reported in [1], the main objective of this OPSYS-subtask is the development of a user friendly software package to support active thermal solar system design, adapted for use with micro-computers. The importance of such computational design tools, the different levels of simulation programs and the specific use of each type of simulation program is described in [3]. The presently available simulation software, of which preliminary versions were presented at the contractor's meeting of november 1986, is briefly described hereafter.

The programs for personal computer systems, as developed in this Opsys-subtask, assume the following type of equipment to be available:
- memory capacity: 512 kbyte or more
- operating system: MS-DOS, version 2.1 or later
- mathematical coprocessor (processor 8087 or 80287)
- 360 or 1200 kbytes diskette drive
- preferably a 10 or 20 Mbyte hard disc, for easy on-line access to various files used by the program during the interactive procedures.
- graphical display with a 640 x 200 pixels resolution.

The equipment described above is frequently encountered in most design offices and consultancies.

3.a. EMGP2: a modular simulation program package for personal computers.

A personal computer version of the transient modular simulation program EMGP2 has been developed and is now available on diskettes, together with a input file generation program and various files to support the interactive procedure. These products are briefly described below.

As compared to the previously available main frame version of EMGP2, the output of the program has been reformatted to at most 80 characters per line. The program interactively requests the names of the input files to be used and displays comprehensive error messages and the requested output on the CRT as the calculation proceeds, irrespective of other output requests.

Because modular simulation programs offer full flexibility to simulate any system which can be decomposed in elementary system components for which an appropriate model, the type of interaction with other components, the applicable control and regulation criteria and the load algorithms are available in the simulation program, the user should be well acquainted with all the program capabilities and its limitations. From this point of view, EMGP2 can not be made user friendly in the sence of taking over this task from the user who is always the only one to know how the system to be simulated looks like and operates. The use of EMGP2 is therefore considered to be somewhat a specialist task requiring a sufficient level of modelling expertise and understanding of the applicability of the available elementary components to the type of system being considered. The required degree of detail in the representation of the various system components and their interaction, in order to simulate the real system operation with sufficient precision for system design purposes, depends in fact very strongly on the type of system and its operating conditions. These decisions can not be made by a simulation program, unless very time consuming sophisticated methods to solve the fundamental multi-dimensional momentum and energy equations would be implemented in the

program. EMGP2 is however based on point models for each elementary
component and idealised controllers, which are to be combined by the
user according to his judgement with regard to the required number of
such elementary components and the size of each element. Guidelines for
the representation of systems by EMGP2-elements can of course be
obtained from the system models used in successfull validation
calculations of the various SPTF-systems considered in the validation
subtask.

In order to facilitate the use of EMGP2, a separate user friendly
interactive program, called EMGP2P, has been developed. This program
serves as an interface between the solar system designer and the
simulation program. It strongly reduces the amount of data to be
specified by the user and generates an appropriatly formatted input file
for the EMGP2-simulation of the user specified system., according to the
selected system representation, specified component properties, control
criteria and load algorithms to be used. The program is based on a
tree-structure of menu's and tables of input parameters, in which the
user can easily find his way to each item to be included in the system
configuration. EMGP2P is supported by comprehensive help files which
are easily accessible at any level of the interactive procedure. These
help files include explanatory texts on all available EMGP2-components,
subsystem schemes, program procedures and physical parameters
encountered when gradually generating the EMGP2 input file. Whenever
detectable errors or data inconsistencies are entered, a detailed error
message together with a suggestion for corrective action, is displayed.
The following block diagram shows the overall concept of the Opsys
EMGP2-software package and the various data transfers involved. This
block diagram refers to the following files which are part of the
software package:
- EMGP2.EXE : executable version of the simulation program EMGP2
- EMGP2P.EXE: executable version of the interactive program EMGP2P for
                interactive input file generation.
- EMGP2.ERR : error message file, used by EMGP2 and EMGP2P.
- EMGP2P.DAT: data file with default values of all parameters.
                The set of default parameters can be replaced by user
                specified parameters by means of the data-save option.
- EMGP2P.HL1 and EMGP2P.HL2: help text files for EMGP2P.
- METFIxxx.TRY: test reference year files, with hourly data.

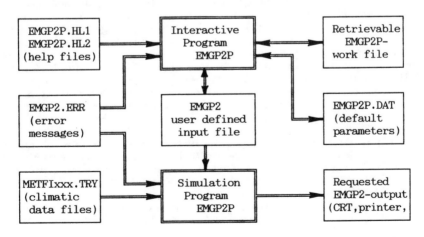

3.b. EURSOL: system specific interactive program for personal computers.

The system specific simulation program Eursol is based on EMGP2 models, but does not require any modelling expertise from the user. The user should however decide what type of system from the series of implemented system configurations, applies best to the solar system for which performance calculations are to be performed. The preliminary version of Eursol presented at the previous contractor's meeting, has been extended with additional system configurations. The program is fully interactive, with a tree-structured menu for the selection of the system to be simulated, supported by an interactively accessible comprehensive help file and uses the same hourly meteorological data files as used by EMGP2. The following system configurations are available; configurations marked with (*) are not yet implemented:

1. Solar water heater configurations: 10 different models are available, according to the following combinations:
- Type of collector loop: forced circulation or thermosyphon
- Type of auxiliary heating: - series connected once through heater,
  - series connected auxiliary tank and
  - immersion heater in solar storage tank.
  (for stratified models only)
- Simulation model used: unstratified or stratified tank model.
  In the stratified tank model configuration the collector array is also represented by a three segment model.

2. Space heating systems and combined systems for space heating and hot water production:
- Type of space heating algorithm:
  - continuous heating at constant flow rate
  - continuous heating with variable flow (*)
  - double thermostat on-off controlled heating
- Type of auxiliary space heating system: series connected or parallel.
  For systems with a continuous heat distribution system with constant flow rate, only the series connected auxiliary heating system can be used.
- Type of solar preheating for hot water production:
  - submerged preheating tank in solar storage
  - external preheating tank in secondary loop.
- Type of auxiliary heating for hot water production: same options as for solar water heaters. For systems with a submerged preheating tank an immersion heater in an external extension of the preheating tank is available as an additional option.
- Simulation model: with unstratified or stratified tank model.
Not all combinations of the above options are meaningfull. The number of available simulation configurations is limited accordingly and the system selection menu will be restructured to account for all meaningfull combinations only.

The present version of Eursol includes a set of minimum and maximum expected values for each parameter to be specified. Parameter values are compared with these limits and a warning message is displayes during the interactive input procedure, whenever a parameter value is specified which is found to be out of this normal design range. The warning message can be ignored or the considered parameter can be corrected.

The solar radiation data processor of Eursol calculates the hourly
values of the incident irradiance for the specified collector
orientation. The results of this calculation can optionally be saved in
a file in order to save computation time for any later simulation
related to the same location and collector orientation.

Although Eursol is on the average about eight times faster than the
modular program EMGP2, the simulation of a combined system with a
stratified storage model may still take 30 to 60 minutes on a typical
micro-computer, depending on the user specified system parameters. To
avoid such long waiting times before a next system can be selected or
parameters can be modified for a next system simulation, Eursol includes
an input file generation option, writing all answers to the interactive
procedure to a user specified file, for all simulations to be performed,
without any calculation at that time. Using this generated input answer
file by means of the MS-DOS input device redirection command, all the
specified system simulations can then be performed, including the
requested output, without any additional intervention. Using this
option, simulation calculations can be performed, for instance during
nights or weekends.

The output of Eursol consists of a detailed monthly energy balance
of the simulated system, on printer or in a user specified file.
Optionally, the complete descriptive input parameter set, a block
diagram of the system energy balance and the monthly histograms of
various output quantities can be added to the output. During the
calculation, results are also displayed on the interactive CRT,
according to one of the following options:
- a summary table of the most important energy values to which a display
  of the instantaneous values of the collector and storage tank
  temperatures can be added,
- a week by week graph of the storage temperature when an unstratified
  tank model is used, or the top and bottom layer temperature when a
  stratified tank model is used. During the calculation, the graphical
  display can be cancelled to return to the summary table of results.

4. DEVELOPMENT OF SIMPLIFIED CORRELATION METHODS ESM1 AND ESM2

This subtask is mainly assigned to Ecole des Mines as a
subcontractor of the Opsys concerted action. Other subcontractors
collaborate in the accuracy and applicability studies of simplified
correlations and in the review of the "Handbook for simplified methods
for solar system design".

The software developed for the simplified correlation methods
requires the following hardware: micro-computer with at least 256 kbytes
of memory capacity, an MS-DOS operating system, one disc drive for 360
kbyte flexible discs.

As compared to the simplified correlations included in the first
version of ESM1 (1983), the algorithms have been improved with respect
to the corrective terms accounting for stratification effects and the
effective base temperature. This improved correlation method has been
validated extensively by comparison with the results of Eursol
simulations, for a variety of system configurations, for the whole range
of practical interest of the main system dimensions and for different
climatological conditions. As a result of this study the limits of use
of the correlation, defined as the range of parameters for which the

discrepancy from Eursol results is smaller than 10 %, have been established. Discrepancies larger than 10 % may appear under the following conditions:
- a collector heat loss coefficient to optical efficiency ratio less than 1 W/(m².K).
- piping heat losses larger than 1.5 times the collector heat losses.
- a storage capacity less than 25 liters per m² of collectors
- a storage heat loss coefficient larger than 30 W/K and per cubic meter
- heat emitter thermal capacitance rate less than the building heat loss coefficient.
- a domestic hot water load which exceeds 50 % of the space heating load
- when high solar fractions in sunny climates are considered: for solar fractions over 50 % or in climates with an average daily total radiation larger than 4.2 kWh/m².

In terms of the collector surface area and the storage volume, the regio range for which solar fractions larger than 50 % can be expected is shown in figure 17. In the final version of this software package, these conditions will be checked and a warning message will be displayed when any of these accuracy conditions would be violated.

The ESM program will be supported by a comprehensive help file, system scheme and by appropriately formatted data files for the monthly cumulative frequency curve parameters for tilted planes and the monthly averaged external temperature, for various European locations. An additional program will enable the user to generate also the CFC-curve parameters for other locations or collector orientations. Output produced by ESM1 will include the specified set of system parameters, the table of monthly results and optionally the sensitivity curves according to the requested systematic variations of system parameters.

For the development of ESM2, a simplified method for medium sized solar water heaters, a very innovative approach is now under development. This new approach is described in [5]. The ESM2-correlation is the result of an analytic integration of the simplified "collector + storage"-equation. The utilisability data are used in the form of typical days. To account for the effect of stratification in the storage tank, the new concept of an equivalent unstratified storage is used. If necessary, to extend the range of applicability, which is now being studied, corrective statistical terms will be defined to improve the agreement with EMGP2-results.

For the "Handbook for solar system design using simplified methods" a draft has been prepared and will now to be reviewed.

References
----

[1] W.L.Dutré: Proceedings of E.C. contractor's meeting held in Brussels on 13 and 14 november 1986, page 149-157 (D.Reidel Publ.Co)
[2] W.L.Dutré: Opsys working document 14: Statistical validation method.
[3] W.L.Dutré: Computational Tools for Active Solar System Design. Proceedings of the European Conference on Architecture, Munich, april 1987, page 349-354
[4] S.Ostergaard-Jensen: Validation of models for solar heating systems. Poster presented at the Ises conference, Hamburg, september 1987.
[5] B.Bourges and J.Adnot: A Generalized Closed-form Model for Solar Hot Water Systems. (Ecole des Mines - Centre d'Energétique)

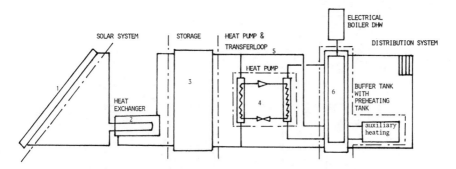

Fig.1 : Scheme of SPTF-SS2-Belgium

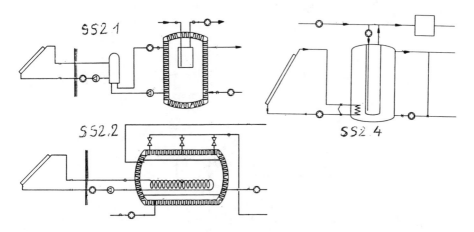

Fig.2 : Scheme of the three configurations of SPTF-SS2-Denmark

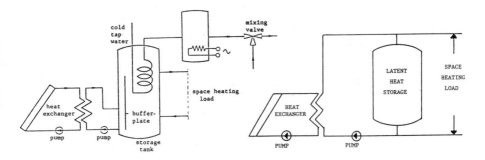

Fig.3 : Scheme of the two configurations of SPTF-SS2-The Netherlands

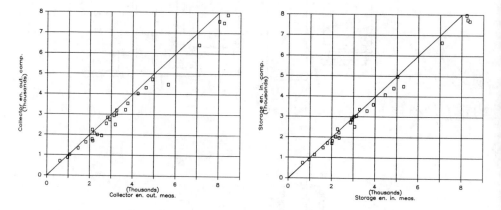

Fig.4 : Collector Energy Output   Fig.5 : Storage Energy Input

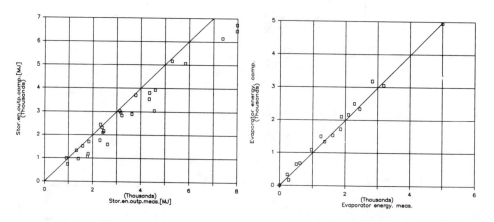

Fig.6 : Total Storage Energy Output   Fig.7 : Heat Pump Evaporator Energy
                                               Input

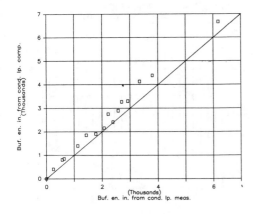

Fig.8 : Secondary Tank Energy Input
from Heat Pump Condensor

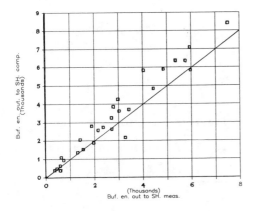

Fig.9 : Secondary Tank Energy Output to Space Heating

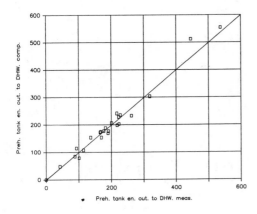

Fig.10: Submerged preheating tank
Energy Output to DHW-load

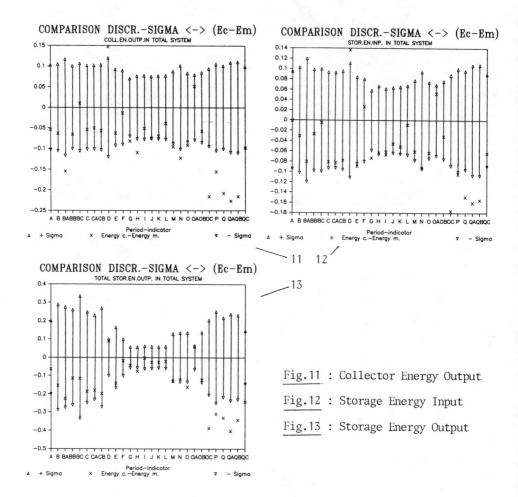

COMPARISON DISCR.-SIGMA <-> (Ec-Em)
COLL.EN.OUTP.IN TOTAL SYSTEM

COMPARISON DISCR.-SIGMA <-> (Ec-Em)
STOR.EN.INP. IN TOTAL SYSTEM

11 12

13

COMPARISON DISCR.-SIGMA <-> (Ec-Em)
TOTAL STOR.EN.OUTP. IN TOTAL SYSTEM

Fig.11 : Collector Energy Output

Fig.12 : Storage Energy Input

Fig.13 : Storage Energy Output

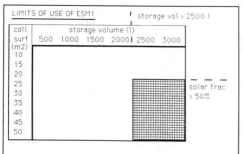

Fig.17 : ESMI1 : Limits of 10 %
accuracy range

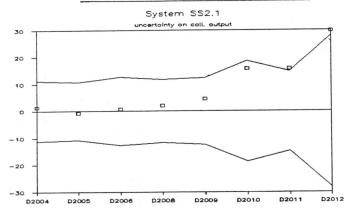

Fig.14 : Collector Energy Output

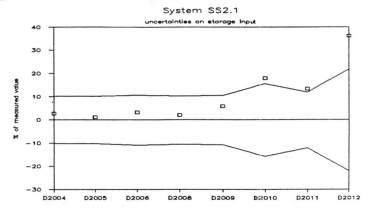

Fig.15 : Storage Energy Input

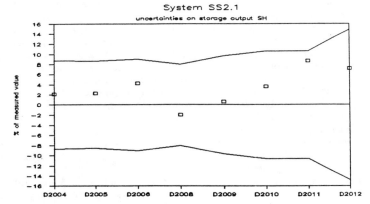

Fig.16 : Storage Energy Output to

THE EUROPEAN SOLAR STORAGE TESTING GROUP
The Co-ordinator's Report on a Concerted Action

Author   : H.A.L. van Dijk and H. Visser

Contract number : EN 35-0045-NL

Duration   : 36 months   from April 1, 1986

Total budget   : Dfl. 3,473,000.--,   CEC contribution: 63%

Head of Project : H.A.L. van Dijk, TNO Institute of Applied Physics
                     (Technisch Physische Dienst)

Contractor   : TNO Institute of Applied Physics
                     (Technisch Physische Dienst)

Address   : P.O. Box 155, 2600 AD DELFT, The Netherlands

Subcontractors : Technical University of Denmark, Lyngby, Denmark,
                     (S. Furbo)
                     Danish Solar Energy Testing Laboratory, Tåstrup,
                     Denmark, (T. Vest Hansen)
                     Ecole Nationale Supérieure des Mines de Paris
                     (jointly with CSTB), Valbonne, France, (P. Achard)
                     University of Stuttgart (FRG), (J. Sohns)
                     University College Cardiff, United Kingdom,
                     (R.H. Marshall)
                     TNO Institute of Applied Physics, Delft,
                     The Netherlands, (J. van der Linden)

Summary

    This paper describes the progress in the European Solar Storage
Testing Group. The aim is to develop procedures both for a simple
test method for the basic characterization of storage systems and
for an extensive test method, resulting in detailed characterization
of the storage systems. The results of the extensive test should
yield parameter values for a calculation model of the storage subjected to the tests. The development of such a general model is a
special task within the research project. The joint action programme
consists of three successive series of tests aimed at evaluation and
improvement of the test procedures and (a) general storage model(s).
A number of subtasks have been defined and distributed to deal with
specific questions.
Many activities are reported of, showing a considerable progress in
the development of a complete, consistent and reliable set of test
procedures. Considerable progress is also reported in the development of (a) storage model(s).
However, the preliminary conclusions presented, have yet to be
verified by series of tests on various types of storage systems. The
evaluation of the first series of (Round Robin) tests has only just
started.

1.   INTRODUCTION

The European Solar Storage Testing Group (SSTG) was established in the beginning of 1982 as a Concerted Action within the second four-year Energy R & D programme (1979-1983) of the Commission of the European Communities. The aim of the Group was to draw up recommendations for European Solar Storage Test methods. The contract ended in 1983 [1] with the preparation of the CEC publication 'Recommendations for European Solar Storage Test Methods' [2], the result of one and a half year of co-ordinated efforts by participants from six European countries.
Within the new EC Non-Nuclear Energy R & D Programme (1985-1988) a follow up of the concerted action on storage testing was started to investigate a number of topics in more detail.
The background, the aim, the scope and the detailed workplan of the project were reported elaborately at the previous contractors meeting in November 1986 [3].

2.   AIM

The aim of the project is to develop a set of test procedures for short term thermal storage systems. The emphasis with respect to the kind of systems and the selected conditions for the tests is on applications in solar energy systems for space heating and/or domestic hot water.
Commercial storage systems often use water as storage material. Nevertheless, also so-called latent heat storage systems are taken into consideration as systems which might be commercialized in the near future.
The aim is to develop procedures both for a simple test method for the basic characterization of storage systems and for an extensive test method, resulting in detailed characterization of the storage systems. The results of the extensive test should yield parameter values for a calculation model of the storage subjected to the tests.

3.   TEST PRINCIPLES

In the Recommendations, published in 1984 [2] procedures are specified to determine the following quantities.:
- the storage capacity as a function of temperature;
- the storage capacity over the design temperature range;
- the heat loss rate at zero flowrate;
- the heat exchanger effectiveness and the potential for
  stratification in the flow direction;
- the storage efficiency.

For this purpose heat is supplied to or extracted from the store during each individual test by means of a fluid with a certain flowrate and temperature.
The flowrate and temperature are controlled in a specific way depending on the aim of the test. An example is presented in figure I. A stepwise change in temperature of the fluid entering the store is followed by a change in temperature of the fluid at the exit. The shape of the curve of this response contains valuable information concerning the characteristics of the store.
In the current SSTG project the possibilities, accuracies and limitations of these and other tests are investigated in order to end with a complete package of test procedures, which meet the requirements defined above.

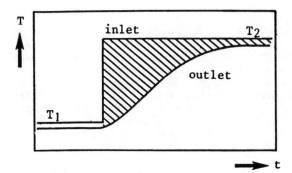

Figure I   Example of a test; the shape of the curve gives valuable information on the characteristics of the store.

4.   WORKPLAN

4.1   Joint actions

The joint action programme is set up around three series of tests
which serve to develop and improve the test procedures and to investigate
the applicability of the test results for the derivation of parameter
values for a computer model of the store.
. Test series A
A Round Robin test on a selected commercially available sensible heat
storage system.
. Test series B
Test on different types of commercially available heat storage systems.
. Test series C
Final test on various types of heat storage systems.
Test series C is follwed by a dynamic test for model validation.

4.2   Model development

Simultaneously, two participants, UK and France, with supporting
effort from the Netherlands, are involved in a special task: the development of one or a few general computer models which can be used to ade-
quately simulate the performance of a thermal storage system.
In this context different possibilities are being investigated.
Towards the end of the project the then available models will be evaluated
and one of the models will be selected and subjected to validation in a
dynamic test. One of the specific requirements of this model is that it
must be compatible with the programme EMGP2 for solar energy systems.

4.3   Subtasks

Eleven subtasks have been defined and distributed among the participants. Each subtask is carried out by a selected number of participants.
Each of the subtasks deals with a specific question, either in relation to
the development of test procedures or the general storage model, or in
relation to possible extensions of the application area.

5. EXPECTED RESULTS

The aim is to conclude the project with the publication of the following final results:
- final recommendations for test procedures for the simple and the extensive tests;
- description of a general method (Fortran package) to process the measured data, including manual;
- description of (a general) storage model(s) for EMGP2, including a clear indication of the possibilities and limitations;
- a list of failures for solar heat storage systems and experienced ways to possibly prevent them.

6. PRESENT STATUS OF THE WORK

6.1 Introduction

Since the start of the new concerted action four meetings have been held:
January 16-17, 1986 in Delft,
May 21-22, 1986 in Delft;
November 3-5, 1986 in Valbonne;
May 11-13, 1987 in Stuttgart.
A number of activities have already been carried out, both within the joint action programme, in the special task on model development and within the various subtasks. Some of the activities have already led to preliminary conclusions. Because of the strong interaction between the different topics under investigation in this project, it is not possible yet to present final conclusions. A certain test may appear very promising in this stage, but might fail when tried out in one of the test series to come, when different types of stores will be tested.
In the following, a number of activities will be presented and discussed.

6.2 Inventory of commercially available storage systems

The inventory of marketed short term storage systems was initiated as one of the activities of the SSTG group with a twofold aim.
The first was to identify the various types of storage systems to which the test procedures should be applicable; the second to select a typical store for the Round Robin test series.
Format sheets were distributed among the participants. Each parti-cipant in the SSTG was assigned the task to organise the collection of relevant information in his country. The emphasis was laid on systems which are known to be applied in solar energy installations, without excluding systems which have a potential to be used in solar projects in the future. The criteria for selection and shifting of the collected data inevitably contain a subjective element. One should, therefore, be careful when comparing the results of the various countries. Table 1 shows some typical preliminary results of the classification of the information received so far.
One should not try to derive any statistical evidence from the data provided. In many cases only one typical example from a series of a marketed store has been put forward. On the other hand series of stores with only a negligible share on the market are represented, alongside widely applied storage types.
The store selected for the Round Robin test series is shown in figure II.

<u>Table 1</u>: Selected preliminary results of the inventory of commercially
available storage systems.

nr. of replies per country	Tank shape:	storage medium:	storage material:
	vertical cyl.:30	water : 37	steel,stainless :10
	rectangular : 4	PCM : 3	steel,glass lining:15
D : 15	horiz. cyl. : 3	pressurized : 30	steel,other or none:8
DK: 14	unknown : 6	non pressur.: 11	copper : 4
F : 3			plastic : 2
NL: 2	<u>heat exchanger type</u>		other + unknown : 9
UK: 6	small helix : 17		
	high vert. helix, small diam.: 2		<u>Main application area</u>
	high vert. helix, large diam.: 6		Domestic HW :33 (15)
	special helix : 2		Space heating: 5 (1)
	mantle : 4		DHW + SH : 6 (2)
	submerged DHW-tank : 2		()=actual application
	special : 4		in <u>solar</u> systems
	PCM-containers : 3		reported

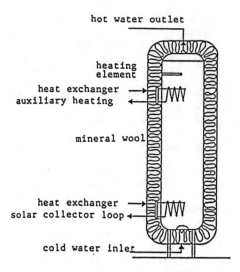

Figure II: The storage system selected for the Round Robin test series.

6.3 <u>Calibration and characterization of the test facilities</u>

Common calibration procedures were defined to evaluate the quality
of the test facilities and the reliability of the results prior to the
start of the first test series.

The main quantities measured during a test are the ambient
temperature, the flowrate, the temperature of the inlet to the store
(either for charging or for discharging the storage system) and the
temperature difference between inlet and outlet (ΔT).

A typical example of the calibration procedure is the measurement of the off-set of the ΔT-probes. The temperature differences are often very small, so special attention is needed for the accuracy of these probes. In this test the temperature difference is measured without a load, both with the ΔT-probe in forward and in reverse direction (figure III). This appeared to be a useful tool to detect and distinguish off-set in the ΔT and the heat loss in the ΔT-probe itself (table 2). The actual test results then can be corrected for these effects.

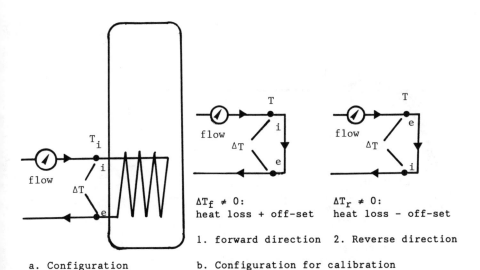

$\Delta T_f \neq 0$:
heat loss + off-set

$\Delta T_r \neq 0$:
heat loss - off-set

1. forward direction   2. Reverse direction

a. Configuration   b. Configuration for calibration
   for testing.

Figure III: Part of the calibration procedure: measurement of the ΔT without a load.

Table 2: Example of ΔT calibration; a typical result (summarized).

	Ti (°C)	V (1/hr)	ΔTf (K)	ΔTr (K)
calibration	70	50	0.112	0.095
		800	0.007	0.002
	45	50	0.044	0.047
		800	0.002	--
Conclusion:	off-set: ΔToff-set ≤ 0.01 K			
	heat loss: Ql' = 0.12 W/K			

Another important quality of the test facilities is the <u>accuracy</u> with which a <u>temperature step</u> can be made. In case of a poor step quality (see figure IV) a correction is needed for the actual energy which is offered to the store. A deviation from an ideal step, however, also affects the response on the outlet, thus making any correction for a poor step not better than a (rough) approximation.

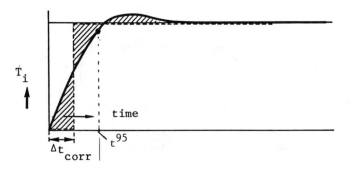

t^{95} : end of step change
Δt_{corr}: translation of time axis as approx. correction.

<u>Figure IV</u>: Example of a poor step quality: quality of the step change in temperature presented at the inlet of the storage.

One of the types of test evaluation techniques currently under investigation uses the transient response of the outlet temperature in the early stage of the test, when the store can still be assumed to be in the initial stage. Such techniques can be of graphical (inclination of curve at certain specific point) or numerical (curve fitting) nature. If in the course of the project such techniques are proven to be valid, then a good quality temperature step is of the utmost importance.

The simplest characterization of the quality of a step is to take the time needed for the inlet temperature to reach 95% of the step, see figure 4. An approximate correction for the derivation of the storage efficiency, taking into account the actual temperature step, can be achieved by translation of the x-axis as shown in figure IV.

To cross-check the accuracy of the total monitoring system of the heat storage test facilities with respect to the measurement of the caloric power which is charged to or discharged from a storage system via the heat transfer fluid it was decided to use a <u>reference heater</u>.
Six identical reference heaters were designed by the Technical University of Denmark, on the basis of heaters already with good experience in use at the university.

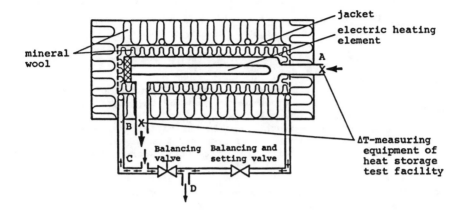

<u>Figure V</u>: Schematical illustration of the reference heater used to cross-check the accuracy of the caloric power measurement.

The reference heater consists of an electric heating element situated in a well insulated tube (see figure V).
The cross-check is carried out by replacing the store by the reference heater. The transfer fluid passing the reference heater is heated up. The electrical power consumption can be directly compared with the caloric power measurement based on the flowrate and ΔT-readings.
The use of the reference heater has led for some laboratories to a reconsideration of the accuracy of their test facility. This is a clear justification of the idea of the cross check.
The reference heater should be used at regular intervals between test series. The advantage of the use of the reference heater is clear: if, in a few tests under widely varying conditions no discrepancy is found between the electric power supply to the reference heater and the heating power measured with flow and ΔT-meters, then confidence is gained in the reliability of the tests, without the need for a full calibration of the test facility.

6.4   Test series A, the Round Robin tests

All (identical) stores for the Round Robin test (figure II) were subjected to a first short test series in the test facility in Delft, before shipping the stores to the other participating laboratories.
At each laboratory identical series of tests have been performed. From the tests, for different temperature levels and different flowrates, the following quantities are derived:

From the step response test:

- the storage capacity as a function of temperature: the amount of energy supplied to the store minus the heat loss during supply;
- the storage efficiency: the amount of energy accumulated within a certain time interval;

From the steady state situation before and/or after a step-response test:

- the heat loss rate at finite flowrate: the heat loss equals the amount
  of energy to be supplied to maintain steady state;

From recharge or from complete discharge after a stand-by period:

- the heat loss during stand-by:
  - recharge : the heat loss roughly equals the amount of energy to be
    supplied after stand-by to recharge the store to its initial
    condition before stand-by;
  - discharge: the heat loss roughly equals the difference between the heat
    content at initial condition and the amount of energy which
    can still be extracted from the store after a stand-by
    period.

The results from these tests are being evaluated at the moment.
The storage capacity and heat loss rates are major parameters in a
physical model of the store. Two other major parameters are the heat
exchanger capacity rate and the amount of (or potential for) thermal
stratification of the store. The latter quantities have a direct effect on
the storage efficiency. However, the storage efficiency itself derived
from the tests cannot be used in a model of the store.

6.5. Thermal stratification and heat exchanger performance

The term thermal stratification refers to the possibility to store
the heat at different temperature levels. In various applications of
storage systems this may lead to an increase of the overall system
efficiency.
It depends on the type of application whether it is important to avoid
mixing (or diffusion) of layers at different temperatures during each of
the characteristic periods of the store: charge, discharge and stand-by.
Tests to measure the (potential for) stratification during stand-by are
still under discussion. The potential to maintain stratification during
charge and discharge tests is revealed by the step response tests, see
figure VI.b.

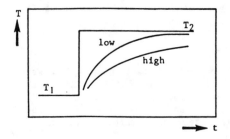 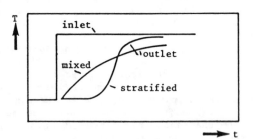

a. High resp. low heat exchanger   b. Mixed resp. thermally stratified
   performance.   storage.

Figure VI: Typical step response curves showing – at least qualitatively –
           the effect of heat exchanger performance and the degree of
           thermal stratification.

The heat exchanger capacity rate quantifies whether the heat exchanger (if present) is capable to transfer the required amount of energy from the transfer fluid to the store (and/or vice versa). Figure VI.a. illustrates how the step response tests also reveal this quality of the heat exchanger.

In the next section it is discussed how the values of these two quantities could be derived from the tests.

6.6 Model development

The work in the area of model development is progressing in a promising way.

At Valbonne, the development of a model identification method is proceeding. The principle fo the technique is to find the mathematical model which gives the best reproduction of the (step) response tests. The results of a few tests have been used to try out different models. It appears at least for the types of store investigated so far - to be possible to create a mathematical model which uses typical physical relations which are also used in the various types of numerical models describing specific storage types.

At Cardiff a model of a 4 port storage system has been developed (figure VII). The model parameters are to be found by fitting the calculated response with the measured temperature curve.

The developments at Cardiff and Valbonne are converging: the new Cardiff model can be considered more or less as one of the mathematical models which the Valbonne technique can produce. The Valbonne technique is more flexible, whereas in the Cardiff model procedures are introduced to accelerate the computation.

In Delft - in close co-operation with Cardiff - series of calculations with the Cardiff model have been carried out, to find the possibilities and shortcomings of the model. Graphical techniques are also still under consideration, for a quick evaluation of the step response curves.

The models will be further developed and tried out using the results of successive test series.

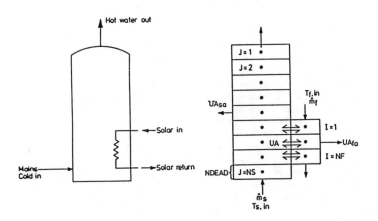

Figure VII: Illustration of the Cardiff 4-port store model.

6.7. Special aspects in numerical modelling

In section 6.5. the term thermal stratification was introduced. Thermal stratification can be destroyed by mixing, dispersion and thermal diffusion.

In a numerical model these three effects can in principle be simulated by an effective thermal conductivity between the successive segments of the store model.

On the other hand, the numerical calculation in itself may lead to a diffusion of heat, at least during the transfer of heat from one segment to the next: in case the time-step for the calculation does not exactly coincide with the fill time of a segment, then the heat in the partially filled segment will be mixed over the whole segment (figure VIII).

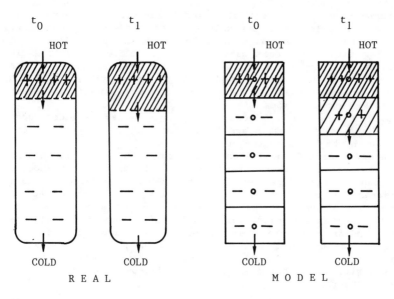

Figure VIII: Illustration of numerical diffusion caused by the calculation model during charge or discharge.

Numerical diffusion could, in principle, be used instead of an effective thermal conductivity to get the calculated step response in agreement with the measured curve. The number of segments is then somehow a measure for the degree of stratification.

At the moment it is investigated how the numerical diffusion depends on the time-step, the number of segments and the type of store, in order either to minimize or to utilize the effect.

One of the complications in this respect is the fact that any solution which would prescribe certain time-step sizes for the calculation would make the storage model incompatible with the EMGP2 model with its autonomous selection of (variable) time-step sizes.

The other complication in this respect is that any solution which would prescribe a certain fixed number of segments, as a function of a.o. the flowrate, cannot be used in case of a variable flow system.

Nevertheless, first calculations are promising, at least for the use of numerical model(s) in the reproduction of step response curves to determine the values for the physical parameters of the store.

And, provided the parameters describe the relation between two physical quantities; they can be offered as input values to any other model which contains the same physical relations and which might be better suited for long term performance assessment of a solar energy system.

6.8. Subtasks

6.8.1 Introduction

The previous report contained the full list of subtasks, distributed among the participants. This report will only highlight the progress within some subtasks. Other subtasks will be reported on in future progress reports.

6.8.2 Subtask 1 (Stuttgart)

A Fortran package has been developed to process the measured data. By means of this program it is possible to evaluate all SSTG charge and discharge tests, heat loss tests at finite flow rate and stand-by heat loss tests. Also a user guide has been prepared. At this moment, an update version of the program with corrections and completions is being used in the evaluation of the Round Robin tests.

6.8.3 Subtaks 2, 3 and 4 (Lyngby and Stuttgart)

Theoretical study led to the determination of a minimum test duration to reach steady state. The conventional criterion for steady state, namely the temperature difference between inlet and oulet of the store ΔT, has been omitted, because in some cases the measuring error in ΔT is too large to detect a still significant - but slow - change in time. For example, in case of a fully mixed 1000 l store heated from 20 °C to 50 °C, with a heat loss rate in the order of 2 W/K it has been calculated that the error in the heat loss is higher than 5 percent unless the test continues until the storage temperature changes approximately 0.002 K per hour. According to the conventional criterion, steady state would have been assumed much earlier, thus resulting in a very inaccurate value for heat loss rate.
The new criterion requires the estimation of the heat loss rate, the heat exchanger effectiveness and other typical thermal characteristics of the store.
The criterion has been tested on two different storage devices: a 1000 l store with a bottom heat exchanger coil and a 20% so called 'dead volume' below the heat exchanger and a 225 l store with a mantle heat exchanger and a small dead volume.
The experiences with both stores were opposite.
For the store with the mantle heat exchanger the calculated stabilization times at the end of a charge test were on the safe side whereas the store with the heat exchanger coil required a much longer stabilization time after charging via the coil, because of the very slow warming up of the zone below the coil. The theoretical analysis of the steady state criterion is being adapted for stores with large dead volumes. However, one has to bear in mind, that for some stores - e.g. with all connected pipes within a bottom 'dead zone' - much larger stabilization times give unrealistic results.

Theoretical investigations of the <u>suitability of different volume flowrates for heat loss tests</u> yielded recommendations for the flowrate in a stand-alone heat loss test.

A high flowrate leads to a low temperature difference, thus increasing the error in ΔT. On the other hand, a low flowrate may not only lead to an increased uncertainty in the flowrate measurement, but also to long stabilization times. Figure IX illustrates for a common heat storage type with heat exchanger coil the optimum flowrate: 1 l/min. This result is supported by experiences from tests on different heat storage systems.

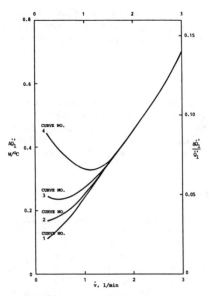

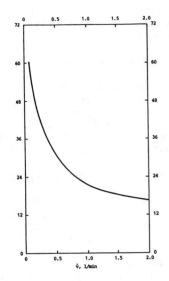

a. Calculated absolute and relative error in measured heat loss rate in case of assumed inaccuracy for ΔT: 0.10 K and for the flowrate: different assumptions (curve 1-4).

b. Required test duration as function of flowrate

<u>Figure IX</u>: Illustration of the optimum flowrate (1 l/min.) for a stand-alone heat loss test.

<u>Three different stand-by heat loss test methods</u> have been tested on the stores which were mentioned before. The results of these tests have been compared with results based on internal temperature measurements.

The test methods with <u>recharge</u> and with <u>direct discharge</u> after the stand-by period gave for the store with the mantle heat exchanger (relative) heat loss values which were in good agreement with the results based on the international measurements. For the store with the large dead volume the heat loss test with recharge gave different steady state situations before and after stand-by (see figure X). So, direct discharge would probably yield more accurate heat loss values than recharge after the stand-by period.

It depends on the shape of the store which of these two tests is the most accurate.
In a third heat loss test method investigated the store with the mantle heat exchanger was <u>completely drained</u>. The method appeared to be very inaccurate, partly because of the large error in the measurement of the temperature of the drained water and partly because of the fact that the drained water temperature is a bad reflection of the heat content of the store. Decrease of the inaccuracy would make this stand-by heat loss test method even more laborious and time consuming than it is already. Therefore, this method has been rejected.

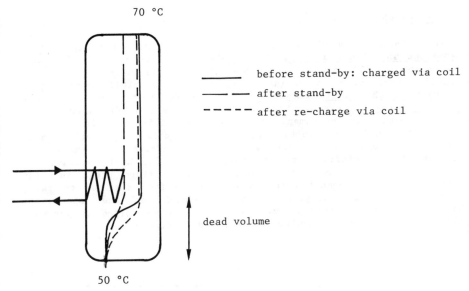

70 °C

```
_____  before stand-by: charged via coil
___ . ___  after stand-by
___ ___ ___  after re-charge via coil
```

dead volume

50 °C

<u>Figure X</u>: An example of the course of temperature in a store with a large dead volume before and after the stand-by period and after recharge.

A detailed model has been used to study <u>the influence of the heat loss distribution over the storage surface on the total heat loss</u> of the store. The model has been applied to simulate heat losses through the container and insulation material of a specific store as well as via thermal bridges.
Investigations on the relationship between the location of thermal bridges and the yearly savings of a specific solar energy system containing the heat storage device showed the importance of the knowledge of the heat loss <u>from the top</u> of the store. This led to a proposal for a test method to determine the heat loss capacity rate from the upper, the middle and the lower part of the heat storage separately. The proposed test is being tried out at the moment.

6.8.4 <u>Subtask 5</u> (Delft)

The measured storage parameters are used in a simulation model to obtain the - e.g. yearly - performance of a solar energy system containing the storage device. Inaccuracies in the storage parameters affect the calculation of the yearly performance.

The influence of measuring errors in the storage parameters on the performance prediction of a domestic hot water system and a combined space heating and domestic hot water system has been examined by means of the EURSOL program (based on EMGP2).
A first series of calculations gave the impression that the influence of measurement errors on the prediction of an annual system performance is not dominant. However, future calculations will focuss on optimized solar energy systems in which uncertainties in the storage performance might have a more distinct influence.

6.8.5.Subtask 11 (Delft)

The principal aim of this subtask is to evaluate a method by which the apparent specific heat capacity of PCMs can be determined. For this purpose a special DTA-instrument has been designed and built.
The special features of this instrument are the possibility of direct measurement of the sample temperature and the realisation of sufficient uniformity of the temperature throughout the large samples. Measurements carried out with this instrument have shown that it is possible to make a clear distinction between melting points and melting trajectories. Side effects like supercooling can also be detected. Furthermore, an alternative method to determine the apparent specific heat capacity has been tested. This method uses a step-wise temperature programme. The results show good agreement with those obtained from experiments using continuous temperature programs. A major drawback of the 'step-wise' method is the long measuring time as well as its limited temperature resolution. An example of results is presented in figure XI.

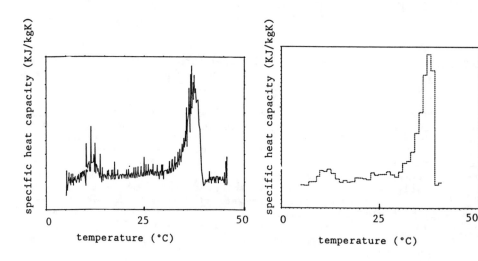

a. with the novel method

b. with the (time-consuming) 'step-wise' method.

Figure XI: The apparent specific heat capacity as a function of temperature for a paraffin mixture.

7.   CONCLUSIONS

    In the previous sections activities were presented, showing
considerable progress towards a complete, consistent and reliable set of
test procedures to characterize the performance of a short term solar
storage system. Considerable progress was also made in the development of
(a) storage model(s) for the derivation of the main thermal
characteristics of a store from the tests and for the simulation of the
storage under realistic dynamic conditions.
    Nevertheless, all preliminary conclusions have yet to be verified
by series of tests on various types of storage systems. The evaluation of
a first common series of (Round Robin) tests has only just started.

8.   REFERENCES

[1]   Galen, E. van.
       The European Solar Storage Testing Group, proceedings of the EC
       Contractors Meeting, held in Brussels, 1-3 June, 1983.
       Project A: Solar Energy Applications to Dwellings. D. Reidel
       Publishing Company, Dordrecht, The Netherlands.

[2]   Galen, E. van and G.J. van den Brink.
       Recommendations for European Solar Storage Test Methods (sensible
       and latent heat storage devices), EUR 9620, November 1984.

[3]   Galen, E. van and H.A.L. van Dijk.
       The European Solar Storage Testing Group, proceedings of the EC
       Contractors meeting, held in Brussels, 13-14 November 1986.
       Volume 2, Solar Energy Applications to Buildings and Solar Radiation
       Data. D. Reidel Publishing Company, Dordrecht, The Netherlands.

" CAM.UR : A COMPUTER AIDED MANAGEMENT SYSTEM FOR URBAN PASSIVE RENEWAL "

Contract number : EN 3S 0048 - B

Duration   : 36 months   1 July 1986 - 30 June 1989

Total Budget   : 764 014 ECU   CEC Contribution : 450 068 ECU

Head of Project   : Dr. A. DUPAGNE

Contractor   : Laboratoire d'Etudes Méthodologiques Architecturales

Address   : Université de Liège
                     15, Avenue des Tilleuls, Bât. D1
                     B - 4000 LIEGE, Belgique.

Summary

By a "general scope of the CAM.UR research programme", we remind of the description of the research context, as well as the statement of the problem and the task description. The "final product description" explains the three parts of the expected results : a computer package for urban passive renewal, an urban reference book and testing actions. The main part of this paper concerns the "state of development". This programme is made up of teams coming from different horizons, some specialized in computer, others in town planning and architecture. Progressing in the development of CAM.UR requires unceasing testing of the new defined concepts by people involved in the everyday practice. That's what we do and what we try to demonstrate in the following pages.

1. GENERAL SCOPE OF RESEARCH

1.1. Description of the research context

In order to reach the best advantage of available passive techniques, it's absolutely necessary to take at the same time into account -the complex differentiation of each building- and the detailed nature of its specific environment. According to the large number of data that are to be manipulated, we couldn't consider the whole town at a time, but only its adequate subdivision into districts, each being constituted of a limited number of blocks. The Urban Improvement Zone (U.I.Z.) - what we called previously Intervention Unit - is, in turn, a part of the examined district : a street with adjacent buildings, a whole block, a place... This hierarchical break up of the town is done manually by the designer according to its own experience. But this difficult task will be made easier by the use of a reference book produced during the research work.

1.2. Statement of the problem

a) The design context

- It is mainly concerned with data acquisition and transformation : climate, site, users, comfort (private and public open spaces), list of possible retrofit actions... It consists of a set of DB with their management modules allowing to transform the row data into the specific ones needed to solve the problem.
- Any design problem being peculiar, it must be carefully negociated with the people concerned. The computer package to be produced will provide the designer with an efficient support in this negociation.

b) The design language

It is a highly interactive system offering site and urban tissue models and 3D descriptions of buildings and public spaces. Three levels of detail are needed for the U.I.Z. representation :

- The first one, very general, is only concerned with global parameters describing the U.I.Z. It refers to knowledge gathered from typological studies and tries to specify the problem and the possible retrofit strategies according to the European region concerned.
- The second one considers the U.I.Z. as a global physical entity. It tries to reach an acceptable equilibrium between private and public requirement satisfaction. All the evaluations are performed on the whole unit and assess the global impact (visual, financial, energy, comfort...) by using simple evaluation methods.
- Once a global rehabilitation strategy or a set of possible retrofit actions have been defined, they are to be transferred to the local physical level of each element contained in the U.I.Z. (individual buildings, public spaces, urban facilities...). Various graphic documents and reports are produced demonstrating the practicability and the interest of the proposed rehabilitation action at the individual, micro-economic point of view.

c) The coordinating system

According to the peculiar design context and to the type of problem to be solved, adapted evaluation tools are to be chosen. These tools give a performance profile for any project contained in the data base. The coordinating system also requires a <u>decision process</u> comparing the performance profile of the project solution to the requirement profile contained in the problem definition. A multicriteria decision process helps the participation team to accept or refuse the proposed solution or to change the problem definition.

1.3. <u>Task description</u>

The present research work intends to make the best use of the knowledge acquired by previous research programmes. It is more concerned with <u>valorization</u> than with developing new fundamental work in the solar energy domain. It intends mainly to produce a large <u>computer framework</u> facilitating the integration of existing softwares in a flexible and easy to use environment. This computer architecture will be built on actual international <u>computer standards</u> (UNIX, relational data base operating systems, GKS...). It will use <u>new hardware technology</u> (powerfull 32 bits graphic workstations, network...) and be open to <u>new software technology</u> (expert systems...).

The main phases of the work to be carried out are :

- functional specification of the whole computer system (main functions and functional architecture)
- global data and context definition (information gathering and structuration, functional performances definition)
- work and data bases (specification and implementation, development of utilities and transformation tools)
- evaluation tool adjustment
    * modules for global energy balance and solar recovery factor at the U.I.Z. level and at the building level
    * thermal comfort evaluation program in public open spaces; temperature analysis and temperature breezes
    * economic viability calculations at U.I.Z. level as well as at micro-economic level
    * visual impact analysis
- output system development
    * external link specification with existing CAD and video systems
    * presentation tools and report generator
    * packaging

Here are the main topics of the different tasks :

Task n° 1 : General coordination

- transmission control
- verification of the outcome
- decision points and possible revision of the research strategy.

Task n° 2 : Functional specification

- specify the main function of the system
- define its functional architecture.

Task n° 3 : Global data and context definition

- gathering of global evaluation methods to define a <u>methodology</u> allowing to approach the <u>urban renewal.</u>

Task n° 4 : Case studies

- feed up the "global Data" task
- define a methodology for urban renewal.

Task n° 5 : Data Base specification

- define the different data bases that are to be used.

Task n° 6 : Data Base transformation tools definition

- specification of the DB tools (management, manipulation, testing,...).

Task n° 7 : Evaluation tools – GLOBAL

- quantitative parameters will be defined in order to give quickly a profile to the existing site and project.

Task n° 8 : Data Base implementation

- computer development of the required Data Bases.

Task n° 9 : Data Base transformation tools implementation

- development of transformation tools.

Task n° 10 : Evaluation tools – DETAILED

- different kind of tools with different levels of detail and technical approach would be necessary.

Task n° 11 : Computer environment specifications

- specifications to be used by the CAM.UR group for the partial standardization of their computer environment.

Task n° 12 : External link

- analysis and if possible specification of the computer links with existing commercial systems.

Task n° 13 : Implementation of the computer environment

Task n° 14 : Presentation tools

- development of programs specially devoted to the presentation of the outcomes of the evaluation tools.

Task n° 15 : Testing

- a tentative application of the package will be tried and tested on an actual city district.

Task N° 16 : Packaging

- detailed description of the produced computer system.

To bring to an end, here is the programme timetable :

CEC-CAM.UR PROGRAMME
"Computer Aided Management for Urban Passive Renewal"

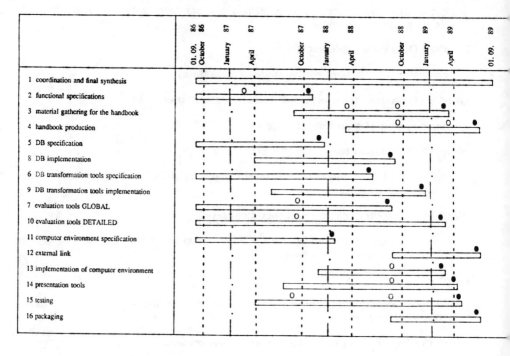

Figure 1. Research programme CAM.UR timetable.

2. FINAL PRODUCT DESCRIPTION

The final product of the research work is threefold.

a) Computer package for urban passive renewal

A computer package for the UNIX environment on graphic workstation will be produced. This computer architecture will be composed of the following items :

* A data base system to store catalogs (ECO, vegetal cover, climate, wall structure...) and the U.I.Z. at different level of detail (geometrical representation, attributes...)
* Tools to input and modify the U.I.Z.: 3D editor, attributes editor
* Tools to create catalogs
* Easy accesses to evaluation programmes (cost, energy...)
* Utility programmes, presentation tools (results of evaluation, visualisation...).

A camera-ready copy of the user's manual will be provided. The LEMA of the University of Liège will, if requested, make the computer package available to European users against payment.

A chapter describing the content of the computer package will be written for insertion into the European Passive Solar Handbook.

This computer package is intended to be a general structure that can easily be adapted and fed up before any commercial distribution to end-users throughout Europe

* It will be completed and examplified for some countries and at least for Belgium
* The package will be supplied with some evaluation tools (ESP, Methode 5000) but its structure will be open to other programmes.

b) Urban reference book

The hierarchical decomposition of the town into districts and U.I.Z. will be based on a reference book. This book will clearly show the link between the intervention unit, the type of urban tissue and the type of urban space. It will also help in presenting reference case studies of successful urban renewal. A camera-ready copy of this book will be provided.

A provisional table of contents of this book is the following :

1. Methodological introduction

2. Case studies of urban renewal performed on typical Intervention Units

- type of climate in EC countries;
- type of urban tissues;
- associated typical urban forms;
- associated typical performance values, indices and parameters;
- associated potential technologies.

3. The project programming tools

3.1. Methodology and organization of all useful information needed to establish the problem.

3.2. Define global objectives and constraints.

3.3. Define boundary conditions (U.I.Z., T.U., D.U. limits).

3.4. Available resources for retrofit.

3.5. The CAM.UR computer package (facilities constraints, evaluation tools, ...) used in the programming stage.

4. The design language

It provides a tool for generating innovative design hypothesis.

4.1. The language is formed by a general grammar structure presenting the "sited type" organization which is the settlement system.

4.2. The CAM.UR computer package used as a 3-D representative system of urban space.

5. Tools for the quality control of the building environmental performances

5.1. Methodology for representation and simulation of the performances of a given project.

5.2. Design hypothesis formulated through graphic representation.

5.3. The CAM.UR computer package as a tool for design evaluation used in the design control stage.

c) Testing actions

A tentative application of the package will be tried on an actual city district by ABACUS and LEMA. For testing, there is a possibility of using one of two settlements in which P. GEOGHEGAN is involved : Limarick City (pop. 80.000) and Westport (pop. 5.000).

A trial application will be carried out by ETHNOKTIMATIKI on a Greek district called Ano Liossia. The following steps describe the work :

- choose definitely a place test (probably Ano Liossia, but possibly Pagrati);
- define the limits of the U.I.Z.;
- gather the data needed by the computer programme;
- definition of potential retrofit actions at the architectural and urban level;
- choose which combination is to be implemented;
- define a detailed action (plot by plot) in the limits of the U.I.Z.;
- produce a report on this peculiar application.

A report of this experience will be produced by the Greek participant.

3.   STATE OF DEVELOPMENT

In this section, we are going to present, in a quite detailed way, the research work carried out by LEMA during this period of the programme PASTOR -CAM.UR. The research work achieved by all the other teams concerned with CAM.UR will be explained by each co-contractor.

3.1.   LEMA research work

Four parts make up this paragraph :

- functional specifications of the CAM.UR computer package;
- specifications of the CAM.UR software;
- CAM.UR DB structure;
- application of the concept.

3.1.1. Functional specifications of the CAM.UR computer package

The CAM.UR package comprizes four entry points, corresponding to the four main action-domains occuring in urban-renewal representation and evaluation :

- assessment of the local resources potential capacities and constraints existing in the examined U.I.Z.;

- evaluation of the global performances of an existing urban design as a whole and of a rehabilitation project, both at the R.Z. level;

- evaluation of the performances of a Decision Unit inserted in an urban context;

- comparison and choice of alternative designs on multicriteria basis.

These four entry points cannot be exclusively considered as a defined set of solving tools, each being specifically devoted to the solution of one peculiar problem, even if some of the actual urban problems can be approached with the help of only one of the entry points.

Things can be largely more complicated and, for example, a single problem can require the use of more entry points than one and eventually all of them. Of course, all the entry points can easily be reached in sequence, each stage giving access to the next one. But one can also get at them in any order or separately according to the nature of the actual problem to be solved.

None of these entry points give access to a rigid and deterministic process. On the contrary, each has a specific menu allowing to adapt the process to the structure of the problem solving as it has been defined outside the computer package.

a)   First entry point : Definition of the design context

This point allows to estimate the rehabilitation potential of a given site. Without developing any definite project, it defines limits to the solution space and ranges of acceptable values for the main parameters characterizing the site.

b)　Second entry point : Evaluation of an urban design

When entering this sub-process, the designer is supposed to have completed a sketch design for the U.I.Z. outside the computer, using the traditional way of design. Of course, he can take advantage of the results obtained eventually by the first sub-process if it has been employed for the problem definition.

This second entry point is made up of two parts :

- first, detailed data acquisition and evaluations of the existing design context at the R.Z. level. The objective is to carry out a more detailed approach of the existing context inside the R.Z., preparing the data acquisition, the evaluation and the comparison of different global sketch plans;

- secondly, test of a project at the R.Z. level. The objective is to encruste a sketch design in the R.Z., and then to carry out evaluations and successive modifications until obtaining an acceptable project.

A 3-D description of both the existing design and the project, as well as their environment allows to carry out numerous evaluation procedures extended to the whole R.Z., applied to inner and outer spaces.

c)　Third entry point : Detailed test of a proposal for a D.U. renewal

This entry point is made up of two parts :

- first, detailed data acquisition and evaluations of a part of an existing D.U. The objective is to carry out data acquisition and evaluations of an existing building (or a set of buildings belonging to the same D.U.) inside the R.Z., exposed to renewal;

- secondly, test of an option of the project at the D.U. level. The objective is to introduce a detailed solution by modification of the building(s) input at the first part of this entry point and successive evaluations.

d)　Fourth entry point : Make a decision

The purpose of this entry point is to help the choice of a project within a group of mutually exclusive candidates. This case occurs when some alternative projects have been developed in parallel or when several variantes of the same projects are available.

A performance profile is calculated for each candidate and a multicriteria analysis is then carried out on the whole in order to rank the projects.

The decision process is based on a programme called PROMETHEE presenting the great advantage of a large choice of models for the preference representation using the concepts of criteria, quasi-criteria and pseudo-criteria. This allows the process to remain as close as possible to the variable nature of the decision context.

3.1.2. Specifications of the CAM.UR software

According to 3.1.1., the CAM.UR software includes five points :

Point 1 : Definition of the design context.

Point 2 : Detailed data acquisition and evaluations of the existing design context at the R.Z. level.

Point 2' : Test of a project at the R.Z. level.

Point 3 : Detailed data acquisition and evaluations of a part of an existing D.U.

Point 3' : Test of an option of the project at the D.U. level.

For each of these points, the schema is the same :

1. Tasks to be done beforehand on paper.

2. Previous data acquisition (for the existing) or simulation of the project by modification of the exising situation (for the project). Data acquisition and simulation are both graphic and alphanumeric.

3. Functional specifications of the evaluations.

3.1.3. Structure of the CAM.UR Data Base

In accordance with these specifications (explained in detail in LEM023 "Specifications of the CAM.UR software"), we defined the following conceptual schema.

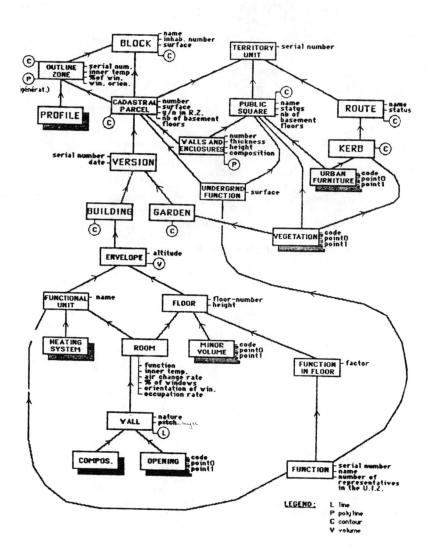

Figure 2. Conceptual schema of the CAM.UR data base.

The specifications and this conceptual schema allowed us to define all the functional dependencies. From these functional dependencies, we established the following 45 relations in fourth Normal Form (4 NF) :

-Uiz Relation-
rUizPt *(UizPt)*

-BLOCK Relation-
rBk *(Bk, BkName, BkSf, InhabNb)*

rBkPt *(Bk, BkPt)*

rCpOutOl *(Bk, CpOutOl)*

-OUTLINE Relation-
rOl *(Bk, Ol, OlTp, OlWindPc, OlWindOr, ProfilCt)*

rOlPt *(Bk, Ol, OlPt, PrOfExLi)*

rOlCp *(Bk, Ol, CplnOl)*

-CADASTRAL PARCEL Relation -
rCp *(Cp, CpSf, CpBsmtNb, CplnRz)*

rCpPt *(Cp, CpPt)*

rCpBsmt *(Cp, CpBsmt, CpBsmtSf, CpBsmtFt)*

rCe *(Cp, Ce, CeThick, CeHeight, CeComp)*

rCePt *(Cp, Ce, CePt)*

rVersion *(Cp, Vers, VersDate)*

rBldgEnv *(Cp, Vers, Bldg, BldgEnv)*

rBldgPt *(Cp, Vers, Bldg, BldgPt)*

rGrdPt *(Cp, Vers, Grd, GrdPt)*

rGrdVePt *(Cp, Vers, Grd, GrdVeCt, GrdVePt0, GrdVePt1)*

-PUBLIC SQUARE Relation -
rPs *(Ps, PsName, PsStatus, PsBsmtNb)*

rPsPt *(Ps, PsPt)*

rPsBsmt *(Ps, PsBsmt, PsBsmtSf, PsBsmtFt)*

rPe *(Ps, Pe, PeThick, PeHeight, PeComp)*

rPePt *(Ps, Pe, PePt)*

rPsVePt *(Ps, PsVeCt, PsVePt0, PsVePt1)*

rPsUfPt *(Ps, PsUfCt, PsUfPt0, PsUfPt1)*

-ROUTE Relation-
rRt *(Rt, RtName, RtStatus)*

rRtPt *(Rt, RtPt)*

rPvPt *(Rt, Pv, PvPt)*

rPvVePt *(Rt, Pv, PvVeCt, PvVePt0, PvVePt1)*

rPvUfPt *(Rt, Pv, PvUfCt, PvUfPt0, PvUfPt1)*

-ENVELOPE Relation-
rEnv *(Env, EnvAlt)*

rFl *(Env, Fl, FlHeight)*

rFlFt *(Env, Fl, FlFt, FlFtPc)*

rFlMvPt *(Env, Fl, MvCt, MvPt0, MvPt1)*

rRo *(Env, Fl, Ro, RoFt, RoTp, RoAir, RoOccup, RowindPc, RoWindOr)*

rWl *(Env, Fl, Ro, Wl, WlTypeCt, WlCompCt, WlPt0, WlPt1)*

rOpen *(Env, Fl, Ro, Wl, OpenCt, OpenPt0, OpenPt1)*

rFu *(Env, Fu, FuFt, HeatCt)*

rFuRoFl *(Env, Fu, Ro, FuRoFl)*

-FUNCTION Relation-
rFt *(Ft, FtName, UizFtNb)*

-TERRITORY UNIT Relation -
rTuCp *(Tu, TuCp)*

rTuRt *(Tu, TuRt)*

rTuPs *(Tu, TuPs)*

-POINT Relation-
rPt *(Pt, X, Y, Z)*

-MOVING FlOW Relation -
rMFw *(MFw, MFwWidth)*

rMFwPt *(MFw, MFwPt)*

Figure 3. Relations in 4NF of the CAM.UR Data Base.

Next, we implemented this relational schema on our SUN workstation using UNIFY Data Base Management Systeme (DBMS).

Now, we develop the software concerning the point 1 of the specifications: definition of the design context. This software, using SUN windowing facilities and user interface facilities as icon, command button, choice button, mouse, ..., will receive plan data from the digitized table, reproduce them on screen, manage them and put them into CAM.UR Data Base.

The next step will be the testing on an existing case coming from Greece or Ireland in order to verify the suitability of the developed software and the reality.

3.1.4. Application of the concepts

The CAM.UR research programme is made up of teams coming from different horizons and specialized in varied domains. Some have good experience in computer, others in town planning and architecture. Ph. GEOGHEGAN and HRU are part of the seconds.

It's fundamental to use the experience of this kind of specialists to suit to the reality. Progressing without references is impossible. The reference information has to exist in the set of information support.

In the way of developing the CAM.UR prototype, it's necessary to surround oneself with people involved in everyday practice. Before going further in the development of the CAM.UR system, the defined concepts had to be tested. So the exercise was undertaken in the application of CAM.UR in Ireland.

The exercise attempts to embrace the different scales of urban renewal projects encountered in Ireland, from village to city scale.

The information is presented as follows :

A. Tralee

The Tralee example is used as an exercise to apply the intentions of the urban intervention Units in a specific location, using the current requirements of the study.

Beside the testing of the concepts, we also show a representation with the required kind of details at the R.Z. level.

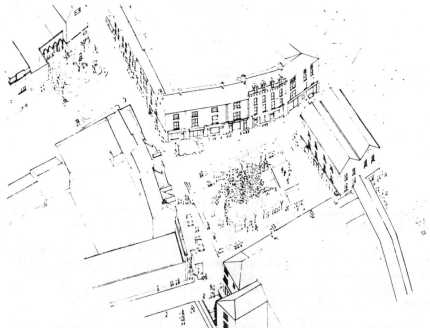

Figure 4. Aerial view of the square.

B. Vicenza, Italy

This example is added illustration of an intervention Unit and a good graphic representation.

All these developments prove that the concepts and the tools of CAM.UR are well suited.

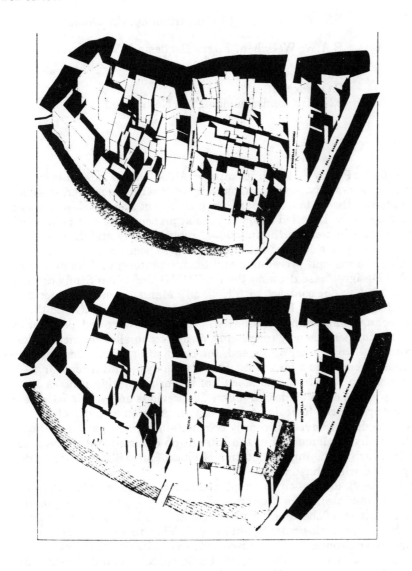

Figure 5. Vicenza : actual state and project.

PASCAUD
Passive Solar in Computer Aided Urban Design

Authors : J. van Dongen, A.W. Tryssenaar, B.A.W. Welschen

Contract number : EN3S-0089-NL(GDF)

Duration : 24 months   1 July 1986-1 July 1988

Total budget : Hfl. 459000   CEC contribution: Hfl. 229500

Head of project : B.A.W. Welschen, Fugro Geodesie B.V.

Address : Fugro Geodesie B.V.
P.O. Box 1213
2260 BE Leidschendam, The Netherlands

Summary

PASCAUD, Passive Solar in Computer Aided Urban Design is carried out as a case-study in the municipality of Haarlemmermeer, Holland. The study adheres to a standard methodology of systems development and is up to the Detailed Design Phase at present. The end-product of this phase will contain the functional requirements and design specifications to a level of detail necessary to begin programming, writing the manual procedures and testing prototypes. The preceding Global Design Phase drew on the PASCAUD Definition Study carried out in 1985 under contract with Holland's Management Office for Energy Research (PEO). Due to the present contract with PEO and the European Commission, the study will be as hardware independent as possible to open the way to systems, owned by smaller municipalities, consulting firms, etc. To this end, the concluding Global Design Report of phase 2 contains a formal information analysis and a resulting "Information Requirements Model" of the urban design process .
In the present detailed design phase the previous analysis is used as a framework for the logical database design. As soon as parts of the database design have sufficiently been detailed, these parts are then individually implemented in the Intergraph environment of the case-study. Much of the implementation of phase four will therefore be done in parallel with phase three.
Present work focuses on the implementation of the "Shadow Chart", which is the interactive projection of shadow polygons in 3D-graphics. The main part of these programs are written in Fortran-77 by Fugro-Geodesie and work presently under VAX-VMS in the Intergraph graphics environment. The general structure however, is system independent, which enables future implementation on different hard ware. The next step in phase 3 is the database design in regard with the chosen (ESP/S) shadow calculation module and the energy correlation model ISSO-16.

PASCAUD, Passive Solar in Computer Aided Urban Design is carried out as a case-study in the municipality of Haarlemmermeer, Holland. The study adheres to SDM, System Development Methodology to manage the WHAT and WHEN of the different subtasks. The main steps in SDM are: 0. Information Planning, 1. Definition Study, 2. Global Design, 3. Detailed Design, 4. Implementation, 5. Installation, 6. Operation & Control. The end-product of the present Detailed Design Phase will contain the functional requirements and design specifications to a level of detail necessary to begin programming, writing the manual procedures and testing prototypes. The preceding Global Design Phase drew on the PASCAUD Definition Study carried out in 1985 under contract with Holland's Management Office for Energy Research (PEO). Due to the present contract with PEO and the European Commission, the study will be as hardware independent as possible to open the way to systems, owned by smaller municipalities, consulting firms, etc.

The method used to structure the contents of the design development is ISAC, Information Systems work & Analysis of Change. In addition to that, a software tool (BLUE-10) is used to analyze and specify the requirements of the participating disciplines in a flexible and communicative way. It enables an extensive and consistent use of ISAC diagrams while it automatically keeps track of the documentation. This means that the transition from generally understood descriptions to a specific computer coding comes only at a very late stage in the process. The resulting modular set-up makes it possible to branch to smaller computer systems without having to start the whole development work all over again.

PHASE PRODUCTS

The body of the Global Design Phase consisted of a sequence of analyses of: the urban design activities, the object area, the precedence of events, and the components & processes. Figure 1 is the resulting diagram of the most aggregated level of PASCAUD, showing only three processes; 1, designing; 2, evaluating the urban and architectural design and 3, evaluating energy aspects in the design. All processes below this level have subsequently been decomposed to such a detail that all relevant components and processes are identified. The information sets and processes to which these sets belong have been described in ISAC diagrams, texts and examples. The result of this forms the contents of the "Information Requirement Model". Parallel to the analysis a framework is made for the selection of (16) models that calculate shading and energy flows. A subsequent assessment has been made for the ten most likely candidates to be implemented in PASCAUD. The end-product of phase two is available in the form of a "Global Design Report".

In the present Detailed Design Phase, the Information Requirements Model of phase two is used as a framework for the logical database design. The end-product will contain the functional requirements and design specifications to a level of detail necessary to begin programming, writing the manual procedures and testing prototypes. The most critical product if this phase will be the physical database design, synthesizing the logical database design, user process requirements and the Intergraph hardware and software of the municipal service of the case study.

The models, selected to calculate shading and energy flows in PASCAUD are ESP Shadow and the Dutch standard correlation model ISSO-16. PASCAUD's alpha/numeric database however, will be made suitable to use the full ESP energy model as an option for individual houses.

SHADOW CHART

Due to time constraints, much of the programming of phase four will be done in parallel with phase three. As soon as parts of phase three have sufficiently been detailed, these parts are then individually implemented in the Intergraph environment of the case-study. Based on phases two and three, a module for the interactive projection of shadow polygons in 3D-graphics was developed by Fugro-Geodesie during the summer of 1987. (Figure 2) The programs are written in Fortran-77 and work presently under VAX-VMS. The input module for this task is written in the User Command Language of the Intergraph environment. The general structure however, is system independent documented, which enables future implementation in different environments.

Fundamental to the shadow module is a vector-based 3D-graphics environment, in which polygons are stored in the design file as a specific type of graphic element and in which polygons are nested in "cells".

The shadow module starts by prompting the operator to accept default values or to enter new values for several parameters (such as the day and time of the shadow projection, the elevation of the reference plane, latitude and difference between local time and solar time). The operator is then prompted to select the object(s) of which the shadow polygon must be generated. Only single polygons, cells or a fence around several design file elements are accepted. When a single element was selected its address in the design file will be stored; when more than one element was selected, an element list will be generated and stored in a separate file; ELLIS. TMP. A Fortran task is then started, in which the direction of the light will be calculated and in which the vertices of the selected polygons are projected on the reference plane. When a single polygon was indicated, the coordinates of the projected vertices are send back directly to the design file as the vertices of the shadow polygon.

When a cell was indicated, with more than one polygon, overlap of the shadow polygon is possible, in which case the borderline of the shadow polygon must be established. In order to do this, the program makes use of a two dimensional byte array (grid), of which the interval (grid-step) and dimensions can be interactively altered. A smaller interval leads to more accurate results, but longer processing times. After all the shadow polygons are placed in the grid array, the borderline is generated by a special procedure, which checks the values of the elements in the byte array. The grid coordinates of the vertices of the borderline are converted to design file coordinates, and the shadow polygon is sent back to the design file. When a error was encountered (e.g. sun too low, grid too small, etc.), a message will be send back to the operator. After the shadow polygon is placed or an error message is send back, the operator can leave the program, or continue by changing the parameters and/or select a new object in the design file.

CONCLUSIONS

The preliminary conclusions of the study are linked with the aim of developing PASCAUD as hardware independent as possible. The complexity of the urban design process makes the use of a proven methodology an absolute condition to reach results within the given budget and time. Especially the information analysis tool BLUE-10 proved to be very helpful in this respect. The number of relations between the defined sets, and the complexity of operations and functions is more than previously accounted for in passive solar research. This underlines both the desired flexibility of the system to be developed, as well as the level of documentation and communication that has to be maintained in the research process itself.

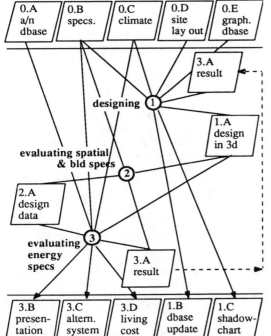

Figuur 1
Precedence diagram of PASCAUD's
most aggregated level

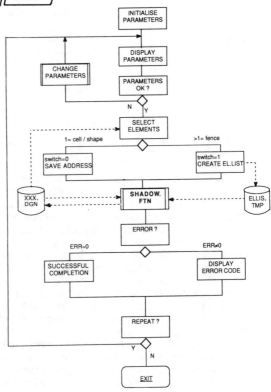

Figure 2. Flow diagram "Shadow Chart"

SURVEY OF THE EXISTING BUILDING STOCK TO ASSESS

THE PASSIVE SOLAR POTENTIAL

Authors   : L. Olsen and O. Olesen

Contract number   : EN3S-0038-DK(B)

Duration   : 24 months   1 May 1986 - 30 April 1988

Total budget   : Dkr. 561.000,-, CEC contribution: 100 %

Head of project   : Vagn Korsgaard

Contractors   : Thermal Insulation Laboratory

Address   : Technical University of Denmark
                     Building 118
                     DK-2800 Lyngby, Denmark

Summary

The aim of the project is to investigate to which extent passive solar energy can be applied in the existing Danish building stock. The first part of the survey is to collect detailed information on a sample of typical building categories. Data concerning plan form, construction, orientation and solar obstructions to the building are being registered. The information is collected by means of on site investigations, deposited plans, ordnance survey maps and various registers. A photographic technique is used for determination of the overshadowing effect. With help of computer and overlay the picture information is transformed into a more comparable form. The data will be analyzed statistically, and characteristics for the building stock will be elaborated. This can be used to define typical buildings which will be the basic of a design study of different passive solar options as : increase of window area, attached sunspace, solar walls, roof-space collectors, increase of building heat storage capacity and reduction of the overshadowing effect. The expected performance will be simulated and the technical and economical feasibility of different designs will be evaluated.

1. Introduction

In the last 10 - 15 years large savings in the energy consumption for heating have been achieved in Denmark. The savings have primarily been obtained by energy conservation measures and energy conscious habits and only with a very limited amount of passive solar measures. In many energy rehabilitated buildings it is difficult to reduce the energy consumption further by energy conservation techniques, but provision of passive solar systems seems to have a large potential for further savings. However, a number of constrains exist for utilization of this technology, and in this study it will be attempted to reveal to which extent passive solar energy can be applied in the existing building stock.

2. Objectives

The aim of the project is to carry out a survey on the existing building stock and thereby identify representative buildings for a design study. Based on these building the most promising passive solar options will be assessed. Many different passive solar measures can be applied to the existing building stock. Some of the most important are :

- Increase of the window area.
- Improvement of windows by, for example, better glazing or movable insulation.
- Attached sunspaces.
- Trombe walls.
- Thermosyphon systems.
- Roof-space collectors.
- Increase of building heat storage capacity.
- Reduction of the overshadowing effect.

3. Selection of sample

In order to fit the general distribution of the building stock in Denmark a number of buildings belonging to different dwelling and age categories have been chosen. The three dwelling categories are :

- Single family dwellings.
- Terraced and low rise, high density dwellings.
- Multistorey dwellings.

The year of construction is divided into 4 categories.

4. Data collection

The information of interest for this project can be divided into three groups :

- Physical data.
- Overshadowing data.
- Energy use.

All the data are collected in standard forms for each building.

The physical data consist of information on :

- Facade
  - area
  - orientation
  - thickness
  - insulation
  - heat storage capacity
- Window
  - area
  - construction
  - shading
- Roof
  - area
  - orientation
  - type
  - slope

- Type and use of rooms behind the facades which are suitable for passive solar measures.

The overshadowing data consist of information on solar obstructions such as neighboring houses and vegetation. The information is gathered for facades and roofs which are suitable for passive solar measures. The most southern facades and roofs are normally chosen for this investigation.

The data on energy use consist of information on :

- Heating
  - system
  - annual cost

- Suggested energy conservation measures
  - type
  - cost
  - annual savings

The information are gathered from :

- Deposited building plans
- Heat inspection reports
- On site inspections
- Photos taken from the facades in question

5. Sensitivity study

The collected data will be analyzed using a standard statistical programme, and characteristics of the existing building stock will be determined. On the basis of these findings, a number of typical buildings will be defined. These representative buildings will be used in a design study in which the most promising passive solar options will be investigated. The expected performances will be simulated, and the technical and economical feasibilities of the proposed designs will be evaluated.

6. The photographic technique

In order to determine the overshadowing effect from neighboring obstacles on a facade or roof a photographic method have been developed.

Fig 1. The camera arrangement :

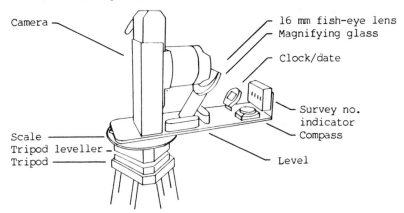

The magnifying glass secures that the clock, the survey no. and the compass are in focus. The lens is a 16 mm fish-eye, f/2.8 which encompas 84° horizontally and 127° vertically.

Fig. 2 Picture taken with camera, with and without overlay :

Procedure for the photography :

A) The camera is placed centrally in front of the facade. It is levelled
   and the orientation scale is set. The height of the camera and the
   distance to the facade are measured.

   Fig 3. The camera set-up :

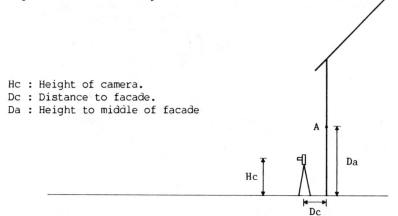

   Hc : Height of camera.
   Dc : Distance to facade.
   Da : Height to middle of facade

B) Three photos are taken; one perpendicular to the facade and the
   others at + 60° and - 60° to this direction.

C) The approximate distances to major objects are recorded.

The transformation from photo to data :

A) The photo is projected down on an overlay chart with a grid. The
   distance between the lines in the grid is 15° horizontally and 5°
   vertically. For each rectangular enclosure the percentage overshadowing is entered in a computer together with the information on
   the distances.

   Fig 4. Overlay :

   Profile :   Percentage overshadowing :

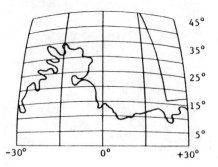

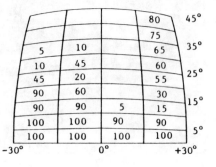

B) The profile from the three photos taken from a facade are put together to one profile encompassing 180° .

Fig 5. Overlay encompassing 180° :

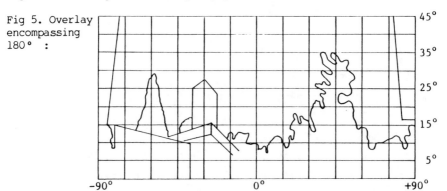

C) The camera height above the ground, the cameras distance to the facade, and the height to middle of the facade are entered into the computer. After a correction the resulting profile indicates the profile from a viewpoint at the middle height of the facade (point A Fig 3.).

Fig 6. Resulting profile :

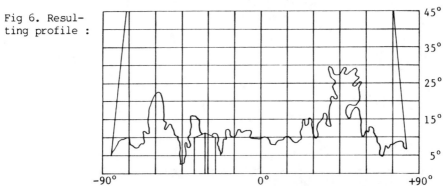

D) From the percentage overshadowing in each enclosure of the resulting profile a factor Fs can be calculated.

The factor Fs is the relative reduction of solar heat in the heating season through a double glazed window in the facade. This factor gives a very good indication of the overshadowing effect on the efficiency of any proposed passive solar measures. The method is described in ref.(1).

7. Conclusions

A procedure for investigation of the existing building stock in order to utilize passive solar energy has been completed. A sample of buildings is now being investigated. The study is expected to be a useful tool for the dissemination of passive solar technology to both the public and the building industry due to the elucidation of the potential of passive solar measures.

8. References

Energiministeriets Solvarmeprogram. Passiv Solvarme - Projekteringsvej-ledning. TI-Varmeteknik, Lab. for Varmeisolering-DtH, Rapp. nr. 30 1985.

PASSIVE BUILDING CONTROL

Contract Number : EN3S-0036-NL

Duration : 36 months   1 Sept. 1986 - 1 Sept. 1989

Total Budget : Dfl. 1417,000.-   CEC Contribution Dfl. 746,000.-

Head of Project : Dr.ir. A.H.C. van Paassen

Contractor : Technical University Delft (TU Delft)

Address : Mechanical Engineering
Mekelweg 2
2628 CD Delft
The Netherlands

SUMMARY

This paper presents the progress of the CEC-concerted action for the design of control systems for passive solar buildings and dwellings.
Each of the three participants has made a "state of the art" report. It turns out that sufficient control devices and sensors are available on the market. However, a lack of knowledge about the characteristics of passive components was observed.
The framework of designs of the control systems are defined and possibilities for the control algorithms are summarized.
TU Delft has built a test cell, so that together with the test cell of ARMINES/CSTB the control systems can be tested.
The activities of the next periods are:
- further development of the control system
- selection of the best control strategy.

1. INTRODUCTION

As stated before in the proceedings of the EC Contractors' Meeting held in Brussels [Paassen, 1986] the project "Passive building control" has the objective to design three prototypes of digital controllers for passive buildings. These systems can manage the passive gains and the heating from the traditional installation in such a way that they minimize the energy consumption and give good thermal- and daylighting conditions all the year round. The participants are the TU Delft (Technical University Delft), ARMINES (École des Mines de Paris) and NAPAC (firm in Paris). The studies of ARMINES will partly be carried out in joint collaboration with CSTB (Centre Scientifique et Technique du Bâtiment). In order to stimulate the exchange of knowledge and to find out what information is available elsewhere, each participant has written a "state of the art" report. A discussion of the results will be given furtheron in the next paragraph. The main difference between the three designs is that the design of the TU Delft is focussed on office buildings and the other participants on dwellings. Although the circumstances in office buildings and dwellings are quite different it may be expected that most of the essential parts will be similar and therefore comparable.

In order to get comparable calculation results a reference model of the building is defined. It will be used for the calculation of the benefits and the control behaviour of the control systems.

It was agreed that all the participants will use the same table of contents in the final report. The advantage of this is, that similar aspects can be found in similar chapters. Moreover, it makes visible all the aspects that should be taken into account. In order to give a quick insight in the control systems a compact characterization will be given of the three control systems. Point by point information will be given about relevant items.

In this paper the latest versions of the three designs are described. Moreover, those aspects are mentioned, that will be investigated in more detail.

The control of the passive systems is rather complicated, mainly due to the lack of knowledge about their control characteristics. Therefore the TU Delft team had decided to build a test cell, although this was not stated in the EC-proposal. This activity was made possible by an extra funding of Dfl. 200,000.- of the Dutch Government (PEO).

With this test cell and that of ARMINES/CSTB measurements can be carried out to obtain a better understanding of the components. Moreover, the prototypes of the control systems can be tested. A description of these cells is given in this paper.

At the end of this paper the research activities as planned for the next period, will be discussed.

2. ACTIVITIES OF THE PAST PERIOD

State of the art

Each participant has written a "state of the art" report. The emphasis was put on the control aspects and the development of the control systems. Worth mentioning are the contributions of ARMINES and CSTB [Archard, 1987]. Although a lot of information was collected it was felt that the reports were far from complete. It was decided to add continuously new information to these reports, so that at the end of the project a concerted "state of the art" can be composed.

Specifications of the reference model

The reference model will be calculated for a simple test cell with a large window facing south. The auxiliary heating is realized with an electric heater. The window has a sun protection system, an insulating shutter, window vents for natural ventilation or a fan for mechanical ventilation. This reference model can be described with state variable equations, such as

$$\dot{X}(t) = FX(t) + BU(t)$$
$$Y(t) = HX(t) \tag{1}$$

F,B,H   are matrices
X(t)   is the vector of state variables
U(t)   is the vector of input variables such as outdoor temperature, solar radiation and heating power
Y(t)   is the vector of the output variables such as the indoor temperature.

The matrices F,B,H apply for a linear system. However, a changing position of the Venetian blinds, shutting rollers etc. changes the system (nonlinear effect). Consequently, the matrices should be calculated for various positions (piecewise linearization). The same applies for ventilation.

Short characterization

Besides the final reports a short characterization will be given. In a form all the relevant items will be compared point by point. Moreover an indication is given where more information can be found. Not all the functions mentioned in the form will be realized during the CEC-project. A lot of functions depend on the hardware and thus on the low cost final configuration, chosen by the firm that will take the controllers in production. This point does not count for the NAPAC-controller, that is already based on a low cost hardware configuration (VISECO).

3. NAPAC CONTROL SYSTEM

The NAPAC control system is focussed on dwellings. Two types of decisions can be distinguished:

a. *Complex and frequent decisions.* The control of the heat gains and/of the auxiliary heat is not simple and not appropriate for the occupant.
b. *Simple and infrequent decisions.* Decisions about ventilation or illumination are simplier to take because they are of the type on/off (window aperture, shading). In actual fact, people have the habit to take such decisions every day. Therefore manual control is chosen. It is proposed to focus on the double control of the auxiliary and of the solar gain. Illumination will be taken into account as a second priority. An advice for the proper manual adjustment of the shading device is given on screen.

System presentation

The proposed control system corresponds to the realistic situation of a room in a solar passive house with electric heating. A dashboard ensures the control of blinds according to the informations provided by the sensors and the user with two criteria: comfort and energy saving. An important feature is the fact that the whole control of the system is not given to the dashboard. The ventilation should not be controlled by the dashboard and the regulation of the blind is user limitated.

The proposed passive system consists of a living room with important solar gains that can be modulated thanks to the blinds. An electric heater ensures the proper indoor temperature. Five sensors give to the dashboard the required information to enable the control of two actuators: the motor of the blinds and the electric heater.

Inputs

The dashboard can accept up to 16 input sensors that should be disposed on the domestic bus. In the simplified system presented above, are considered
- an outdoor temperature sensor
- an indoor temperature sensor
- an electric power sensor
- a light sensor
- a presence sensor.

Each sensor is connected on the domestic bus and has a particular address on this bus: when addressed, it transmits a numerical word that represents the actual value measured (temperature, power, light).

Outputs

The outputs are connected to the dashboard according to the same standard as the inputs; the domestic bus carries the information for output peripherals. The simplified solar system uses two outputs or actuators:
- *A power controller (for electric heater)*. The type of control used is chrono proportional: on/off control with pulse-width modulation.
- *A blind motor*. The blind position is transmitted to the regulation of the motor through the domestic bus.

Control design

The control strategy which is implemented in the main program of the VISECO (third level) uses simultaneously the physical measurements and the user's setpoints. It ensures both the heating system control and the light control; the problem is quite complex because of the conflicting objectives: energy saving and comfort. The control strategy must also be simple enough to be implemented in the dashboard in real time conditions.

To achieve realistic characteristics, the control is elaborated on the following basis:
- A comfort function is given by the user:
  * an indoor temperature setpoint
  * a maximum light setpoint.
    These setpoints can be continuously modulated by the user thanks to the interactive software of the VISECO (second level).
- A thermal model of the dwelling is automatically determined by VISECO during the first days. This model is based upon model identification [Bacot, 1985]. The dynamic model is calculated by the software with the equations (1) of the previous chapter. With this model the best control algorithm is selected.
- The control algorithm minimizes the energy costs acting on the electric power and the blind position, with respect to the two following constraints:
  * nominal light intensity
  * indoor temperature setpoints.

Communications

VISECO has two communication channels:
- The modem channel which ensures the communication with a distant terminal.
- The domestic bus which is reversed to communications between the dashboard and the sensors or actuators.

The dashboard (figure 2)

The dashboard is a compact and stand alone data processor which allows the display of technical informations in the house. VISECO corresponds to technological and ergonomic requirement which makes it a fully userfriendly set.

VISECO has two human interfaces:
- the screen and the keyboard for local programming
- the telephone time interface for distant communications.

The software

As mentioned below, VISECO is running under control of a powerful multi-task operating system.

This software presents the following levels:
- *first level*: control of peripherals, real time clock, keyboard, LCD screen, telephone line (relays, modem, serial access), domestic bus.
  This first level corresponds to very hardware dependent tasks which are continuously running . They can be considered as background tasks.
- *second level*: human interface (menus), physical measurements, storage.
  At this level, the conversation with the user takes place (set points programming).
- *third level*: The main control algorithm takes place at this level; it is running separately from the other control purposes and has the lowest priority. This low priority cannot lead to details greater than one second, so the control algorithm cannot be disturbed by any of all other tasks.

4. TU DELFT DESIGN

The TU Delft Design is focussed on office buildings, that will take into account the changing needs for heating, ventilation and lighting.

System presentation

The auxiliary heating is realized with radiators. Each radiator is provided with a control valve in case of hot water heating or an on/off control device in case of electric heating. The window has a sun protection system (indoor Venetian blinds), an outside insulating shutter and window openings for ventilation.
In each room there is one controller, that is designed as a dashboard giving information to the occupants about the room's thermal state. Moreover, it controls the indoor temperature and/or illumination automatically or by remote control in case the inhabitant overrides the automatic control.

Inputs of the central controller

The input signals will be delivered by a local weather station with:
- outdoor temperatuur sensor
- solar radiation sensor
- wind speed sensor
- wind direction sensor.
The information about the weather forecast of a nearby weather bureau and the information of a local weather predictor [Lute, 1987] may be used.

Inputs of the local room control (dashboard)

Following sensors will be used:
- output of the central computer (desired values)
- indoor temperature
- illumination
- presence of people
- position of the blinds
- position of the window openings.

Outputs of the central controller

The central computer calculates:
- desired indoor temperature for each room
- desired position of window vents' motor
- type of temperature control (by adjusting auxiliary heating, ventilation or passive gains)
- desired position of shutting roller
- desired position of lighting switches.

Outputs of the room controller

The local computer realizes the setpoints as calculated by the central computer and takes into account the overriding actions of the inhabitants.
The outputs are:
- the position of the radiator valve or the electric power
- puls width modulated signal for adjusting the window vents' motor
- idem for the Venetian blinds
- the position of the light switches
- the position of the insulating shutter.

Control system design

The control system consists of a central computer and room controllers. The central computer is the optimiser which provides the best setpoints. These setpoints will be calculated by means of a thermal model of the building and a weather predictor. The thermal model can be identified from the outputs of sensors.
The room controller sets the various components in the desired position and controls the indoor temperature in a close loop configuration. As a first step a prototype that controls a test cell, will be developed and tested. A blockdiagram is represented in figure 3.
No specially designed low cost hardware will be produced. The functions of the local controller for the test cell are carried out by a data-logger with calculation facilities. For the central computer an AT-microcomputer is used. At the end of the project a firm may be interested to convert the arrangement into a commercial version as is demonstrated in figure 4.

5. ARMINES/CSTB DESIGN

The aim is to develop controllers using adaptive strategies for the combined control of direct gains and the auxiliary heater in residence buildings.

System description

The system to be controlled is a one room building based on the test cell. It has a large window facing south and it has a medium range inertia. The cell has an adjustable awning for sun shading and an electrical convector for heating.

Inputs/outputs

The inputs are: the indoor temperature, the outdoor temperature and the solar radiation. The outputs are: the position of the awning and the required electric power of the convector.

Control design

The control strategies that will be studied, range from very simple ones to very sophisticated ones based on the optimal control theory. An example of a simple one is a linear control law that rules the depth of the overhang and the heating power, in accordance with the inside temperature or the outside solar radiation. The output is a proportional, differential and integral function of the input.

A more sophisticated one is an optimal control strategy, based on the minimum principle. It is able to adjust the equipments, taking into account economic criteria and comfort requirements. It uses a simplified thermal model to represent the dynamic behaviour of the building.

The most promising laws will be experimentally tested on the ENSMP real scale test cell. These tests will be carried out through the implementation of the selected control laws in a data-aquisition system. Consequently, no commercial low cost versions will be developed at this stage. Also items as lighting, ventilation are not considered.

Study of control strategies

The study is structured in four main parts:
- The first one concerns the establishment of models of the thermal behaviour of the Sophia Antipolis test cell. With identification methods, models are derived from experimental data files [Gicquel *et al.*, 1987].
- The establishment of control laws.
  * Six intervals of the indoor temperature are distinguished and to each of them a position of the awning is assigned.
  * With a model of the test cell the position of the awning is defined and put in the right position.
- The third part of the study consists of simulations, carried out with the building thermal behaviour detailed simulation code, MIVERVE.
- The last part could be called the practical realization. Each scenario, which parameters have been previously selected, will then be tested in the test cell.

6. TEST CELLS

Test cell of ARMINES/CSTB (figure 5)

The cell is a one room building with a large window facing south. The sizes are $5 \times 5 \times 3$ m (width × length × height). The window of 2.5×2.5 m has an adjustable awning. The heating system consists of a convector, which heat power can be controlled. The walls of the cell are well-insulated. The building has a medium range inertia. It is that height that the main time constant is 38 hours. The heat loss coefficient is $1.8 \ Wm^{-3}K^{-1}$.

Test cell TU Delft (figure 6)

The test cell is a light one. The wall is a sandwich construction, consisting of a polystyreen layer with steel sheets. It is possible to add mass into the room so that the heat capacity of a real building can be simulated. The internal sizes of the cell are $3.10 \times 3.90 \times 2.65$ (width × length × height). The front wall contains a glass facade on a parapet of 0.8 [m].

There are two configurations:
a. - external shutter (k-value = 2.65 [W/m^{2}K])
   - double panes, 5-12-4 [mm] (k-value = 3 [W/m^{2}K])
   - Venetian blinds
b. - external shutter (k-value = 2.65 [W/m^{2}K])
   - double panes (k-value = 3 [W/m^{2}K])
   - Venetian blinds
   - single panes, 5 [mm].

In both cases attention should be paid to the possibility of natural ventilation. In the double glass facade and the single glass facade four small vents can be opened.
In the cell is an electrical heater. The maximum capacity of this heater is 3500 [W]. The lighting consists of 4 TL, each 36 [W]. The Venetian blinds, vents' openings and the external shutter are motorized and can be controlled.

7. ACTIVITIES FOR THE NEXT PERIOD

The activities are focussed on two items:
- *Framework of the control system.* It will be defined in more details. So far only the general structures are given. The control system should deal with a lot of aspects. It is unlikely that all these aspects can be studied during the project's period. Limitations should be indicated and choices should be made.
- *Control strategies.* The main activity will be the study of the optimal control strategies that will deliver the best setpoints for the local control loops. This study will be carried out by computer simulation based on the reference model. Also the control behaviour and the tuning of some of the control loops will be investigated.

REFERENCES

- PAASSEN, A.H.C. VAN, Digital control of passive solar systems. Solar Energy Applications to Buildings and Solar Radiation Data. Proceedings of the EC Contractors' Meeting held in Brussels, Belgium, November 1986, pp. 73-81.
- ACHARD, P., TANTOT, M., Direct gain control using shading devices. ECA, München (1987).
- ARCHARD, P., COOLS, C., Passive solar components using rotating blades including a phase change material, ECA, München, April (1987).
- BACOT, P., Identification de modèles de comportement de systèmes thermique. Revue Générale de Thermique, no. 277, Paris (1985).
- LUTE P.J., and PAASSEN, A.H.C. VAN, Short term prediction of indoor temperature in buildings. European Conference on Architecture, München, April (1987).
- GICQUEL, R., NEIREA, NOGARET, Identification methods for optimal control of heating systems E.C.A., München, April (1987).

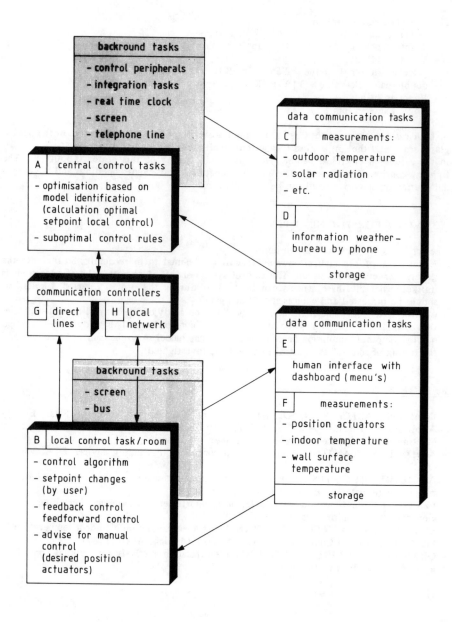

Figure 1: Framework of the control system.

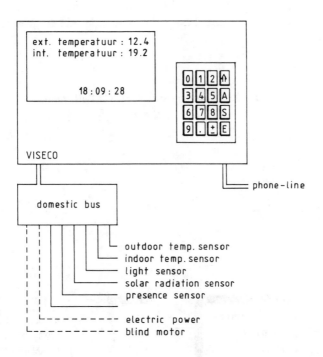

Figure 2: The domestic dashboard and its connections.

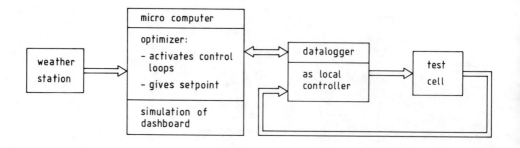

Figure 3: Laboratory prototype TU Delft design.

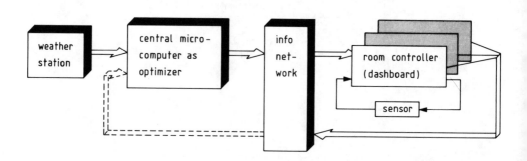

Figure 4: Commercial version TU Delft design (to be designed by a firm).

Figure 5: Test cell ARMINES/CSTB.

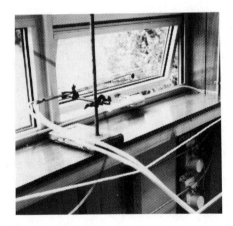

Figure 6: Test cell TU Delft.

INTERZONE AIRFLOW

Contract Number: EN 3 S - 0037 - UK

Duration: 24 months 1-4-86 to 1-4-88

Total Budget: £88,000

Head of Project: John Littler

Authors: Saffa Riffat and Mashhour Eid

Contractor: The Polytechnic of Central London

Address: Research in Building Group
Polytechnic of Central London
35 Marylebone Road
London NW1 5LS

SUMMARY

In this paper we describe the development of a multi-tracer gas system for measuring interzonal air movement in buildings. The system consists of simple and stand-alone gas injectors and sampling units. The injectors are capable of releasing up to four perfluorocarbon tracer gases in different parts of a building. Following tracer gas injection and mixing, small samples of air are collected using automatic sampling units. Each unit consists of a 16 position valve and a group of removable stainless steel tubes packed with solid absorbent. Samples may be taken simultaneously at up to four points in space and up to 16 points in time. These samples are then stored for subsequent analysis in the laboratory using an automatic thermal desorber and gas chromatograph.

1. INTRODUCTION

One aspect of energy conservation attracting attention at present is the reduction of heat losses in buildings derived from poor thermal insulation and high air infiltration rates. As a result a large number of superinsulated houses are being built in Scandinavia, North America and recently in the UK[1]. These houses are constructed in such a way that air leakage through cracks and openings no longer serve as a source of ventilation and so mechanical ventilation systems are required. Inadequate air change rates give rise to an increase in indoor air contaminants (eg, formaldehyde, nitrogen dioxide and moisture) which have an important influence on the health and comfort of the building's occupants. Research is therefore required to evaluate the extent of air ventilation and dispersion of interior contaminants so that the optimum compromise between energy consumption and sufficient air change to maintain a healthy environment is achieved.

The air tightness of buildings can be examined by pressurisation using blower doors. Unfortunately, this technique is unable to provide information on air infiltration in a building under natural climatic conditions. As an alternative, tracer gas techniques may be used for measuring air change rates in buildings. Reviews of various tracer gases and measuring techniques have been made by Harrje et. al[2], Sherman et. al[3], and Lagus and Persily[4]. Until recently, measurement of air movement has usually involved the assumption that the building is a single uniformly mixed zone. However, multizonal measurements are essential if movement of indoor air contaminants and energy transfer among various parts of the building is to be accurately represented. The measurement of interzonal air flows in buildings, such as hospitals, is important as the transport of odours and bacteria between various ward units must be kept to minimum[5].

The use of a multi-tracer gas, instead of a conventional single gas method, increases the speed and accuracy of interzonal air flow measurements[6]. It also clarifies specific flow directions. Gases which have been used include carbon dioxide, nitrous oxide, hydrogen, methane, ethane, helium, perfluorocarbons (PFTs) and sulphur hexafluoride. We have used perfluorocarbons and sulphur hexafluoride in our work as they have desirable tracer gas characteristics in terms of detectability, safety and cost and they have been used successfully in air infiltration studies[6,7,8,9]. Both SF_6 and PFTs exist only at very low background concentrations in the ambient and are easily detected (parts per billion range for SF_6 and parts per trillion range for PFTs) using an electron capture detector.

2. DEVELOPMENT OF A MULTI-TRACER GAS SYSTEM

Experimental work on interzonal air movement was previously carried out by Prior et. al[10] at PCL using an unrefined system consisting of automatic injection and sampling systems. During the present work this system has been improved with the development of the following:

2.1 Tracer Gas Injection

The rate of decay technique is used in our work as it requires relatively simple apparatus.

A number of injection units are being built. These simply consist of a 10 ml aluminium cylinder wrapped with a band heater and fitted with a sealing cap, Figure 1. Prior to each experiment, cylinders are injected with a known amount of perfluorocarbons and sealed in the laboratory. A programmable timer is used to energize the band heater during the test allowing tracer gases to be released simultaneously into a building. The following perfluorocarbons have been used in our work: perfluoro-n-hexane, perfluoro-methyl-cyclohexane, perfluoro-dimethyl-cyclohexane, and perfluoro-decalin.

2.2 The Sampling System

The microcomputer-sampling system developed by Prior et. al[11], may be improved with the following benefits: (1) greater speed of deployment within buildings, (2) a more compact and flexible system so it can be used during periods of a building occupancy, (3) reduced capital cost of the system.

The design of a compact and stand-alone sampling system is shown in Figure 2. The sampler consists of a 16-position, 34-port valve (type ST flowpath), made by Valco Instruments. The valve has an inlet port, an outlet port and a pair of ports at each of the 16 sampling positions. A 16-position version of this valve and a sampling tube is shown in Figure 3. A small removable stainless steel tube, Figure 4, packed with a divinyl-benzene/styrene co-polymer adsorbent is connected to each pair of valve ports. Air at a constant flow rate may be drawn in each sampling point using a small pump and the multi-position valve is positioned by a stepping motor. A two channel digital timer is used as a control system to operate the stepping motor and pump.

The operation procedure of the system is as follows. At the beginning of each experiment the valve is rotated to position 1 and the pump is turned on. At the end of the desired sampling collection time, set by timer, the pump is turned off and the valve is rotated to position 2. This procedure is repeated until all samples have been taken. The system is flexible since the sample loops may be filled in any desired sequence and at any desired time. The samples may be then stored for subsequent analysis in the laboratory.

Work has been completed on building one sampling unit and we are currently building further units of a similar design.

2.3 Tracer Gas Separation and Analysis

A new gas chromatograph, type 8420, made by Perkin Elmer is currently being commissioned. The chromatograph, which is fitted with an electron capture detector (ECD) and an ATD-50 thermal desorber, will be used for separation and analysis of the samples.

3. MEASUREMENTS AND SIMULATION MODELS

The new multi-tracer gas system will be used for measuring air movement in a number of recently constructed houses in the Milton Keynes area. Measurement will be carried out in other passive solar heated houses at Newham, UK. Some measurements may be carried out in Central Europe, providing suitable projects are located.

In the second phase of our programme, a comparison will be drawn between measurements of interzonal air flow and rates predicted using rules of thumb such as those developed by Balcomb et. al[12], and using simulation models such as BREEZE and ESP[13]. In order to make

comparisons with these models, work will begin on the implementation
of the latest versions of the codes, at PCL, in the near future.

4. REFERENCES

1. RUYSSEVELT P. LITTLER J, and CLEGG P.
    "Experience of a year monitoring four superinsulated houses",
    Conference on Superinsulation, Published by UK-ISES, 1987, pp76-89.
2. HARRJE D J, GROT R. A, GRIMSRUD D.T.
    "Air infiltration site measurement techniques", Contribution to the
    2nd AIC Conference on Building Design for Minimum Air Infiltration,
    Sweeden 1981.
3. SHERMAN M. H, GRIMSRU D.T., CONDON P. E. and SMITH B. T.
    "Air infiltration measurement techniques", Proceedings of 1st AIC
    Conference, Air Infiltration Instrumentation and Measuring
    Techniques, UK, 1980.
4. LAGUS P, PERSILY A. K.
    "A review of tracer-gas techniques for measuring airflow in
    buildings", ASHRAE Trans. 91, Part 2, 1985.
5. FOORD N, and LIDWELL D. M.
    "Airborne infection in a fully air conditioned hospitals", J Hyg.
    Camb. 75, 1975.
6. DIETZ R. N. and CROTE E.
    "Air infiltration measurements in a home using a convenient
    perfluorocarbon tracer gas technique", Environment International 8,
    1982, pp419-433.
7. DIETZ R.N., GOODRICH R. W. , CORTE E. A. and WIESER R. F.
    "Detailed description and performance of a passive perfluorocarbon
    tracer system for building ventilation and exchange measurements",
    Presented at the Symposium on Measured Air Leakage Performance of
    Buildings, American Society for Testing and Materials, USA, 1984.
8. HARRJE D. T, GADSBY K, LINTERIS G.
    "Sampling for air exchange rates in a variety of buildings", ASHRAE
    Trans. 88, 1982.
9. Proceedings of 1st AIC Conference, Air Infiltration Instrumentation
    and Measuring Techniques, UK, 1980.
10. PRIOR J. J., MARTIN C. J., LITTLER J. G. F.
    "An automated multi-tracer method for following interzonal air
    movement", Paper HI-85 No 2, Presented at the 1985 annual meeting
    of ASHRAE, Honolulu, Hawaii, 1985.
11. PRIOR J. J. and LITTLER J. G. F.
    "A multi-tracer gas method for following interzonal air movement
    and its application in solar heated buildings", Proceedings of
    the 1st international symposium on ventilation for contaminant
    control, 1985, pp275-289.
12. BALCOMB J. D., YAMAGUCHI K.
    "Heat distribution by natural convection", 8th National Passive
    Solar Conference", 1983.
13. McLEAN D. J.
    "Simultaneous dynamic simulation of air flow and energy in
    buildings", CIB 5th International Symposium.

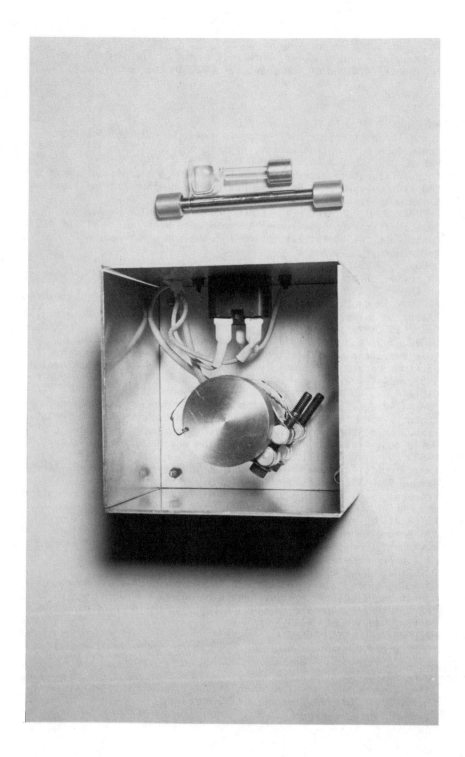

Figure 1   The tracer gas injection system.

Figure 2   The sampling system.

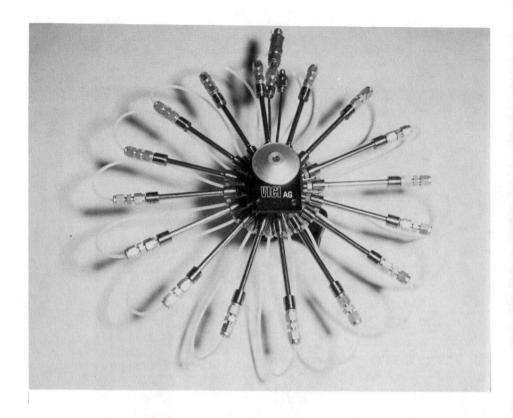

Figure 3   The 16-position valve and the sampling tubes.

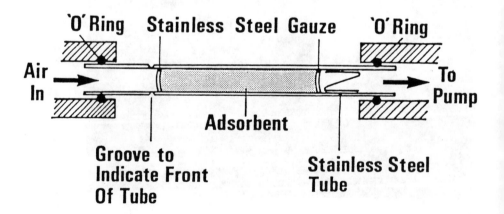

Figure 4   A single sampling tube.

THERMAL COMFORT IN PASSIVE SOLAR BUILDINGS

Authors   :   P.O.Fanger and H.N.Knudsen

Contract Number   :   EN3S-0035-DK(B)

Duration   :   36 months   1 May 1986 - 1 May 1989

Total Budget   :   2,400,000 DKR   CEC Contribution 2,400,000 DKR

Head of Project   :   Prof. P.O.Fanger

Contractor   :   Technical University of Denmark (Laboratory of
                       Heating and Air Conditioning)

Address   :   Building 402
                       Technical University of Denmark
                       DK-2800 Lyngby

Summary

Current thermal comfort models and standards are premised on conditions of thermal equilibrium between the building occupants and their indoor climate. Such standards are therefore not intended for use in passive solar buildings where indoor climatic conditions vary both spatially and temporally. The current report outlines the progress of a research programme being conducted in Denmark, Great Britain and France with the intention of developing a non-steady-state model of thermal comfort. In Denmark experiments with human subjects experiencing operative temperature step-changes have been conducted, and experiments with operative temperature profiles are now planned. The British subcontract has designed a research methodology, and a number of schools, offices, hospitals and dwelling-houses have been chosen for data collection, which is now well-advanced and continuing. The French subcontract identified the thermal behaviour of passive solar buildings by means of a literature review and computer simulation program.

1. Introduction

The detailed knowledge that has evolved in the last 60 years on the subject of thermal comfort has largely been based on the assumption of steady-state thermal exchanges between the human body and its environment. For example, the recent international standard on thermal comfort, ISO 7730 assumes that thermal equilibrium conditions apply to the building occupants (1). The heat-balance models underlying such comfort standards have been based on the results of exposures of subjects in climate chambers to fixed thermal conditions for several hours (2). Models and standards are therefore applicable only to approximately steady-state conditions. Some buildings have significant solar gains but may still maintain approximately constant thermal conditions. For such passive solar buildings the existing standard can be applied.

But many passive solar buildings are typically employing thermal storage of the structure to take advantage of the solar radiation, when available. The indoor climate of such passive solar buildings is characterised by significant spatial and temporal variations in operative temperature. It is the purpose of the present research to establish thermal comfort criteria during such transient conditions.

The current paper outlines the progress of a) a basic laboratory study of human response to thermal transients (Danish project); b) a field study of thermal transient and human comfort within real occupied passive solar heated buildings (British sub-project); and c) a computer simulation of the thermal performance of passive solar buildings using meteorological inputs from different climatic locations within Europe (French sub-project). The core of this research programme is the model of transient thermal comfort that is expected from part a, which will be field-validated in part b, and subsequently used to predict human responses to the indoor thermal conditions being output from the passive building thermal performance simulations in part c.

2. Basic Laboratory Study of Transient Thermal Comfort (in Denmark)

2.1 Introduction

The laboratory study of transient thermal comfort consists of two series of experiments:
- operative temperature step-changes
- operative temperature profiles.

The former serie of experiments has been performed and the method used, and some preliminary results will be described. The analysis of the results and the planning of the latter series of experiments are now in progress.

2.2. Method for the Operative Temperature Step-Change Experiments

In two identical climate chambers separated from each other by an air-sealed door subject experiments were carried out to investigate the thermal response to operative temperature step-changes. The partial vapour pressure and the mean air velocity were kept identical in both chambers.

Twelve subjects were selected for the experiments. They were split into two teams of six, and each experiment was conducted twice.

Physiological measurements such as skin temperatures according to the procedure outlined by Olesen (3) and rectal temperature at a depth of 8 cm were recorded every 5 minutes except for the first 5 minutes after the step-change where the frequency was once every minute.

The subjects were asked to cast a thermal sensation vote (ASHRAE scale) every 5 minutes throughout the full three hour-experiment. But during the first 5 minutes after the step-change they were asked to vote every minute starting with an immediate vote when entering the second chamber.

During the experiments three levels of clothing insulation were studied: nude, 0.6 clo and 1.0 clo. The two clothing ensembles were made of polyester, a non-hygroscopic, man-made textile fiber. This material was chosen to minimize the absorption and desorption effects in the clothing due to the differences in the relative humidity in the two chambers. At each level of insulation four different experiments were conducted leading to a total of 12 experiments. During all experiments the thermal conditions were chosen to obtain a neutral thermal sensation either before the step-change or some time after the step-change. The other conditions were chosen to be warmer or cooler than neutral so that the following experiments were conducted: neutral to warm and vice versa, neutral to cool and vice versa.

2.3 Results

The analysis of the results is now in progress. Preliminary examples of experimental results are given in Fig. I and Fig. II which show the mean thermal sensation vote for the twelve subjects during two of the total of twelve experiments where they were dressed in 0.6 clo polyester and experiencing operative temperature step-changes of 4.5K. It is seen that the downstep in thermal sensation was larger than the upstep, and that the transient was felt immediately, that is, the new level for thermal sensation was reached just after the change. These findings are general as long as experiments near neutral or cool are concerned.

2.4 Operative Temperature Profiles

From the French subcontract some representative indoor temperature profiles occurring in passive solar houses will be chosen and used for experiments in the climate chamber, to investigate the thermal response of subjects under such conditions. These studies are now being planned.

3. Field Studies of Passive Solar Buildings and their Occupants (British Subcontract)

3.1 Introduction

Two important features differing between conventional architectural spaces and passive solar environments are the greater possibilities for change in thermal conditions over space and time. The well-established theories of thermal comfort do not allow us to predict how human beings will respond to the two forms of transient which could arise from these changes: fluctuations in temperature which come about either because the person moves from one micro-environment to another, or because the micro-environment surrounding a stationary person changes.

3.2 Method

In parallel with the major Danish laboratory study of transients we are carrying out a number of case studies intended to cast light on these problems by generating subjective data (instantaneous assessments of thermal sensation and comfort at the beginning and end of transients, post-evaluations of the experience of specific changes, generalized longer-term evaluations of the importance of changes in conditions in the buildings occupied) and objective data (instantaneous assessment of clothing and activity levels, physical measurements of thermal conditions at the time of assessment, longer-term descriptors of variability within the occupied buildings). This work is to be carried out in a range of building types and, while most fieldwork will be carried out in the UK, confirmatory studies will also be conducted in the Federal Republic of Germany and France.

Statistical analyses of these data are intended to allow the construction of empirical models of response to change which can be compared with the laboratory results.

3.3 Results

The design of an appropriate research methodology has now been complete for some time and a number of schools, offices, hospitals and dwelling-houses in the three countries have allowed us access for data collection, which is now well-advanced and continuing. Since levels of cooperation from building users have been lower than expected sampling is not yet complete.

The database is being augmented as information is collected but data analysis has not yet commenced and therefore no quantitative findings can yet be reported. However, at least as far as solar houses are concerned, the researchers' initial impressions based upon interviews with residents are that variation over space seems to be more important than variation over time, and that in multi-storey houses differences between ground and first-floor conditions are more important than differences withing floor levels. We have had the opportunity to compare owner-occupied solar houses with housing-association solar houses and it appears that there are no clear thermal differences between the two samples of residents that might relate to the owners having selected a solar home (in fact the solar and energy-related features of the houses do not appear to be especially important in the house-buyers' decision-making). A small number of those cooperating with us had either designed or made a substantial contribution to the design of their houses and some of these had designed variation in, as a desirable feature. Spatial variation is also seen as a positive feature by some lay building users, since it offers the possibility of comfort at a variety of activity levels.

It should be stressed that these are very preliminary impressions and must await quantitative verification.

4. Identification of the Thermal Behaviour of Passive Solar Buildings (French subcontract)

4.1. Introduction

The thermal behaviour of passive solar buildings has been identified in two ways:
- a computer simulation of characteristic passive solar buildings,
- literature survey covering experimental studies.

4.2 Computer Simulation

4.2.1 Introduction

This part of the French work is now finished. The results are presented in the second volume of the French June 1987 progress report. The purpose of that first part of the French contribution in the "Thermal Comfort in Passive Solar Buildings" project was to identify typical operative temperature swings in passive dwellings.

Two kinds of buildings can be defined: (a) Classical passive solar buildings, i.e. direct gain buildings. Their south oriented glazing area was either little (only 4 m^2) or quite important (about 16 m^2), b) Passive solar buildings with specific wall like "Trombe wall"; the area of that wall was 6.8 m².

In order to define important variations of operative temperature, the summer period was considered more particularly. Simulation during autumn, spring or winter period did not show important amplitudes of indoor temperatures (no high overheating).

The French study was based on a real passive solar house built in the northeast of Lyon by two architects named G.Aymard and M.Goy (see Fig. III).

4.2.2 Method

The previous mentioned house has been exposed to four climates: Copenhagen, Kew, Paris and Rome. Climatic data originated from Test Reference Year tapes. Two climatic sequences were selected per site: spring period (about March or April depending on the site), summer period (July month for all the sites).

4.2.3 Results

The results are given by the mean air temperature versus time, the mean radiant temperature versus time, the relative air humidity versus time, the relative mean air velocity versus time.

An indication, called "quality mark", based on the international standard, identified whether or not the user was in a comfortable situation. Some samples are given in Figs. IV, V and VI. We can notice 3°C for radiant temperature amplitude and 5°C for air temperature amplitude. Amplitudes of these magnitudes may imply incontestable discomfort. It is interesting to notice that the level of natural ventilation rate influences this problem, depending on the level of outdoor air temperature.

Solutions for such problems are important during the design of buildings, particularly in solar protection of glazing areas.

4.3 Literature Survey

4.3.1 Introduction

The aim is to provide real scenarios of "indoor cliamte" in passive solar houses.

The knowledge of the "indoor climate" needs the determination of the environmental parameters previously mentioned. All these parameters are necessary to define the user comfort.

4.3.2 Results

The study has shown that most of instrumented houses have few measuring points and that motivations of measurements concern energy rather than thermal comfort.

So, wall temperatures, for example, as well as air velocity were almost never measured.

The conclusions are:
- In most examined projects, the only measured parameter is the air temperature (available in data files form or diagrams form).
- Air velocity and air humidity were never measured in occupied houses.
- Mean radiant temperature was rarely measured. In a few cases, wall temperatures were available for one, two or three surfaces.

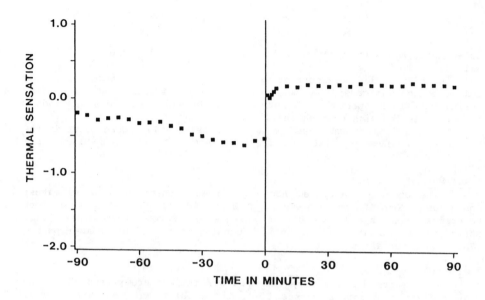

Fig. I: Mean thermal sensation for subjects wearing
          0.6 clo polyester and experiencing an upward
          operative temperature step-change of 4.5K.

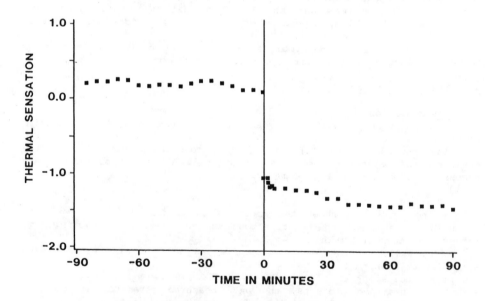

Fig. II: Mean thermal sensation for subjects wearing
          0.6 clo polyester and experiencing a downward
          operative temperature step-change of 4.5K.

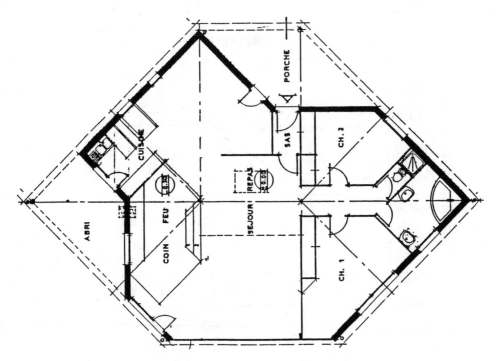

Fig. III: Map of the project. Dagneux house (France).

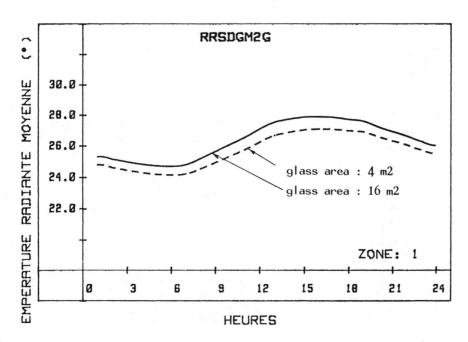

Fig. IV: Variations of mean radiant temperature for two glazing areas
(July TRY, Rome).

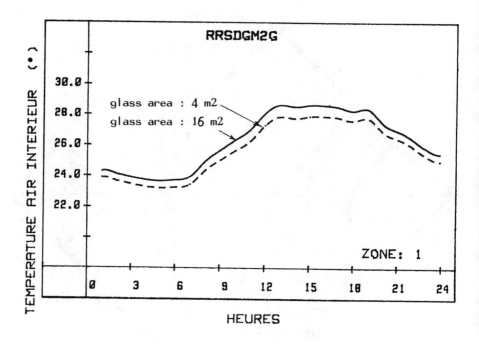

Fig. V: Variation of air temperature for two glazing areas
            (July TRY, Rome).

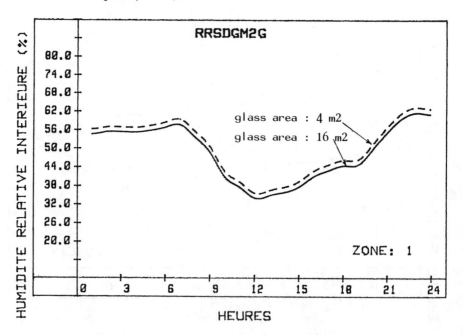

Fig. VI: Variation of air humidity for two glazing areas
            (July TRY, Rome).

References

1) ISO (International Standards Organization) (1984) *ISO 7730 : Moderate Thermal Environments - Determination of the PMV and PPD Indices and Specification of the Conditions for Thermal Comfort* (Geneva: ISO).

2) Fanger, P.O. (1970) *Thermal Comfort : Analysis and Applications in Environmental Engineering* (Copenhagen: Danish Technical Press).

3) Olesen, B.W. (1984) "How many sites are necessary to estimate a mean skin temperature?" In *Thermal Physiology*. Ed. Hales, J.R.S. (New York: Raven Press), pp.33-38.

FIELD-BASED RESEARCH ON THERMAL COMFORT IN PASSIVE SOLAR BUILDINGS

Contract Number	:	EN3S-0090-UK
Duration	:	23 months 1 October 1986 - 31 August 1988
Total Budget	:	£101 260 CEC Contribution £101 260
Head of Project	:	Mr I D Griffiths, Department of Psychology, University of Surrey
(Co-workers	:	Mr A P Baillie, Mr J W Huber, Dr R Samuels)
Contractor	:	Department of Psychology, University of Surrey
Address	:	University of Surrey GUILDFORD Surrey GU2 5XH UK

Summary

In recent years there has been a move in building design away from regarding solar radiation as a problem and towards using solar energy as a positive contribution to the reduction of the energy demand of buildings. Most European and international comfort standards are based on a very well-established biophysical theory of thermal comfort, which finds its empirical foundation in laboratory studies but which has specified higher comfort temperatures than have been evidenced by field studies. It has also so far provided little information about variations in the environment and the effects of solar radiation. It has also stressed optimisation of thermal sensation rather than the satisfaction of human comfort requirements per se. This all has obvious implications for the application of established theory to new forms of building, such as passive solar environments.   Our methodology is based upon the statistical analysis of standardised interview data and simultaneous physical measurements, carried out in a sample of houses, offices, schools and hospitals. These are selected in such a way as to encompass both a number of buildings with passive solar features (primary focus) and a control group of non-solar buildings.

Analysis of results from non-solar buildings shows that comfort temperatures are dependent upon context, and not only upon clothing insulation and activity level. Qualitative impressions (of a totally interim nature) are that it may well be the case that building occupants, even those who have selected passive solar buildings, do not primarily assess those buildings in terms of their thermal or energy-related characteristics.

1.1 Introduction

In recent years there has been a move in building design away from regarding solar radiation as a problem and towards using solar energy as a positive contribution to the reduction of the energy demand of buildings. Since in Western Europe as a whole the greatest energy use is the provision of comfortable conditions inside buildings it is necessary to keep comfort standards under constant review.

Most European and international comfort standards are based on a very well-established biophysical theory of thermal comfort, which finds its empirical foundation in laboratory studies of how people report their sensations during exposure to different sets of conditions in controlled environment chambers.

In practice the laboratory studies have almost always led to the prescription of higher comfort temperatures than seem to be required in occupied buildings. There is so far no convincing explanation of this difference, although factors such as possibly higher activity levels in real life rather than laboratory simulations, and social psychological factors such as attitudes, knowledge and beliefs, have been suggested as having a role here. If indications that the context or setting is of significance in determining which temperatures are comfortable are proved correct, then it may be that the biophysical formulae will need the addition of further terms for building type and type of occupier. This has obvious implications for the application of established theory to new forms of building, such as passive solar environments.

In addition laboratory work has in the past emphasised the study of steady-state conditions, although many passive solar features are characterised by variable conditions, either in terms of differences between the south side of the building and the north (which will be experienced as people move about), or, in the sun-spaces themselves, there may well be transients which arise from, for instance, shading of the building.

Laboratory studies tend to be directed at theoretically defined optimal conditions (for instance, thermal balance or the absence of thermal sensation), rather than satisfaction or comfort per se, and thus we have aimed to develop techniques for the assessment of these neglected factors.

Laboratory studies are not able to simulate the experience of solar radiation, which because of its psychological significance (particularly to people in the more northern latitudes) may well produce human reactions which are not predictable from information about reactions to other forms of thermal radiation.

Finally, thermal comfort, while an obviously important criterion in assessing the success of buildings with passive solar features, is not necessarily more important than other factors such as daylight, sunlight and aesthetics, which may interact to produce the overall level of satisfaction with the environment within a building. In this case, the fieldwork approach may yield results which are directly relevant to the designer's task: the imaginative solution of a number of problems simultaneously.

1.2 Method

Our method is based upon scheduled interviews carried out with the occupiers of selected buildings, which are accompanied by simultaneous measurements of salient features of the thermal environment. Since the interviews yield quantitative evaluations of the building users' experiences together with data about activity and clothing insulation levels, statistical analysis will provide results which are directly comparable with biophysical theories. In addition psychological data are gathered about a wider range of building features () so that the interactions of those features which have significance to the users can be investigated. Both psychological and physical measurement programmes contain elements which are specifically targeted at the investigation of transients and spatial non-uniformities.

The study requires the application of novel techniques of data collection and therefore a programme of pilot studies, principally of offices and homes, is required to ensure that interviews and measurements are optimised for our purposes.

We are selecting a sample of buildings for investigation which is intended to cover a broad range of building types: homes, schools, offices and hospitals. While, clearly, the focus of sample selection is upon buildings with passive solar features it is necessary to include a smaller sample of conventional buildings as a control sample, since we believe that reactions to the same physical conditions may be different in buildings designed according to different, and evidently different, philosophies. While there is a bias within our sample in that most of our effort is concentrated in Great Britain, smaller-scale investigations are also being carried out in France and the Federal Republic of Germany.

Buildings and their occupants are investigated over the seasons, but, practically, it is not possible to investigate all buildings in all seasons, and thus the sample changes in constitution as the study progresses.

1.3 Results

We have now completed all phases of the piloting of research techniques and are fully occupied in fieldwork. As data are collected they are processed to form part of an expanding database. It would be an inappropriate use of resources to carry out a number of interim statistical analyses and therefore it is only possible at present (with the exception of a small number of pilot studies within conventional buildings) to report which buildings have been investigated and to report on the impressions of the research workers who have been carrying out the studies, rather than any completed analysis. It should thus be made very clear that these results are subject to later revision and can only be treated as temporary indications.

Pilot studies have been limited to Great Britain and have involved an office building and a college of technology with atria, two conventional office buildings with unusually large glazed areas, and a number of passive solar homes. Quantitative investigations have been carried out at two passive solar housing developments (at Milton Keynes and Chorley), at a passive solar office and a college of technology at Farnborough and a shopping mall atrium in Surrey. Three non-solar office buildings in London and Surrey have been investigated, and studies of a solar hospital in London and a control hospital in Kent are about to start, together with an experimental low-energy office with some solar features (in Hertfordshire). Negotiations are in progress for access to solar and control schools in Hampshire.

In France cooperation has been secured at passive solar houses in the NW and SW of France and at 2 schools in the SW. These have already been investigated in the early summer. In Germany an investigation has taken place in a group of solar houses in Freiburg and it is planned to carry out a study of some solar houses in Ingoldstadt. Preliminary contacts have also been made with some occupiers of office buildings.

1.4 Analysis of results and comments

The UK Department of Energy's Energy Technology Support Unit funded, as preparation for the present study, a large part of the pilot studies. For practical reasons the quantitative aspects of these were carried out in non-passive-solar buildings but, because of innovations in methodology, have already produced results which are of interest in this field; notably the finding that while biophysical comfort theory provides a sound account of human reaction to thermal conditions in offices the model does not seem to explain people's thermal responses when at home.

Since no real analysis of results in passive solar buildings has yet taken place all that it is possible to present here is some early and impressionistic indications. While these suggest the sort of conclusions which we may arrive at they need to be treated with considerable caution.

It is fairly clear from the housing studies that houses are not selected on the basis of solar features alone, or primarily. The sun space is frequently seen as a garden feature and there is often the belief that the house is unusual in concept, or has been specifically designed, is of importance in itself. The French sample is different in this respect since a larger proportion of the occupants has been directly involved in the design. There is often a degree of dissatisfaction related to either excess expectations (e.g. that energy bills will be even lower than they in fact were, or that sun-spaces would be inhabitable for both plants and humans in all seasons) or due to quality of building factors related more to cost than to concept. In low-cost British solar houses the northern side of the house often suffered from condensation and mould growth caused by low temperatures and low ventilation. Open-plan living spaces were positively valued in the houses studied in France, but not in the UK and Germany where the non-compatibility of various activities was often seen as more important than the functional flexibility afforded by the layout. The use of open spaces as air-circulation voids frequently led to dissatisfaction, either in temperature or air-movement terms. There were sometimes unfavourable daylighting implications arising from the presence of plants in the sun-spaces.

A similar problem arose in the schools investigated, where teachers found the limitation of daylight which was involved in controlling solar overheating was adversely commented upon. Teachers were also conscious that they lacked control skills and information and resented the diversion of their efforts from teaching which was involved.

In the passive solar office investigated (Briarcliff House, Farnborough) occupants placed greater emphasis on the internal appearance of the building than the internal environment conceived in thermal terms: the light-well and internal garden were sources of pleasure, but the external canopy was seen as unpleasant, and there was also a lighting problem in that internal illumination levels were set at values which matched conditions when solar-control blinds were closed. It was in support of a totally passive approach that automatic solar control devices had had to be modified so as to operate more slowly or immobilised because their operation distracted users from their work.

It has been noted that the degree of cooperation with our research from building occupiers has been lower than previously encountered. This is particularly true of commercial and institutional occupiers and has resulted in a considerably increased effort being required in securing access to any individual building. Some of this is undoubtedly due to the fact that many of the buildings which could play a part in the study are already ´over-researched´, and some to changes in the general social climate.

2 Conclusions

At the time of publication we are part-way through a field-based study of thermal comfort in passive solar buildings, which is intended to parallel and complement laboratory-based studies of thermal comfort. Emphasis is currently placed on the collection of relevant physical and psychological data and little analysis has so far been carried out.

Quantitative findings indicate strongly that human thermal requirements are not uniform once activity level and clothing insulation are taken into account, but rather depend upon the wider environment.

Qualitative impressions seem to indicate that the inhabitants of solar houses are not primarily motivated by their enthusiasm for the principles of solar architecture, but do perhaps like the ´unusualness´ of their house. Their assessments of the house seem not primarily to be made in terms of energy or the thermal environment. Analogous situations seem to apply in schools and offices, where positive features which arise as by-products of passive solar design (sun-spaces, light wells) are evaluated favourably in their own right. Negative features (extra effort, loss of daylight) are also seen as individual factors, rather than in the context of the whole design.

Gaining access to solar and control buildings has proved more difficult than anticipated, and further effort is being directed to meeting this need.

EUROPEAN CONCERTED ACTION PROGRAMME ON DAYLIGHTING

Contract Number: EN35S-0047-I

Duration   : 36 months   1 September 1986 - 31 August 1989

Total Budget   : 932.400 ECU   CEC Contribution: 649.700 ECU

Head of Project: Prof. Aldo Fanchiotti, University of Rome

Contractor   : ICIE - Istituto Cooperativo per l'Innovazione

Address   : ICIE
                   Via Nomentana 133
                   00161 Rome, Italy

Summary

The outline of the three-year concerted action programme
on the use of natural light in office buildings is
recalled. Six research groups of five countries are
involved in the programme. The main objective of the
programme is the promotion of the use of daylighting
concepts and materials by designers, particularly as far
as office and schools building are concerned.
The main results obtained in the course of the first year
of activity are briefly presented.
These include: a. study of algorithms to derive luminous
data from energetic data; b. selection of algorithms to
estimate illuminance and irradiance on any surface; c.
critical review of new materials; d. classification of
components; e. study on control strategies and sensors,
as well as on artificial lighting equipment; f.
identification and analysis of high quality "daylit"
buildings; g. review of numerical simulation and design
tools.

1. Introduction

Until recent years, most of the effort for promoting energy saving and solar energy applications in buildings has been limited to the thermal aspects: passive solar or bioclimatic architecture.

Daylighting is rapidly becoming one of the most interesting items in the area of energy saving in the building sector, particularly as far as commercial buildings, with predominantly diurnal occupation, are concerned.

Architects appear to be much more interested in, and sensible to, the luminous, rather than the thermal aspects of solar radiation, as the problem of the relationship of buildings with light is a traditional concern for designers since the very beginning of architecture. Light is what makes a space visible and colours and textures perceptible.

Thus, daylighting might very well be the key for gaining the attention of designers for energy saving principles and technologies.

Unfortunately, while numerous examples of brilliant natural lighting applications can be noted in the work of many architects, both of the past and of the present, the average, current commercial buildings design practice very seldom reflects a conscious approach to the energy saving implications of using daylighting concepts.

The objective of the European Concerted Action Programme is to promote, among designers, the use of natural light as a source of light for commercial buildings, looking contextually at the thermal and luminous aspects of solar radiation, by means of appropriate information and prediction design tools.

2. Research Areas

The basic problem can be described as follows: how to effectively and economically bring natural light to parts of a large commercial building that are located far from the building envelope, as to replace artificial lighting, so saving energy and creating better conditions.

This research programme is based on the identification of the many sub-problems to be solved in order to deal with the main question, which led to the definition of the following research areas /1/:
a. Daylight availability data: data necessary as input for simulation or design tools;
b. Luminance models: algorithms to describe complex configurations;
c. Human comfort and response: human response to light proper-

ties and to illuminance levels;

d. Materials: innovative materials with advanced optical properties;
e. Components: light transmitting, screening, redirecting and distributing devices;
f. Controls: hardware and software for measuring and controlling light levels;
g. Artificial lighting equipment: innovative lamps and fittings compatible with natural lighting;
h. Systems: innovative design building, making use of new materials, components, control, equipment;
i. Numerical simulation: codes for simulating thermal and luminous aspects;
j. Design tools: to be used by designers at the different stages of the design process.
k. Economics: economic implications of using daylighting techniques.

Some of the most interesting achievements of the Programme in different research areas in the first year of its implementation are presented in the following chapters.

3. Participants

The following six groups are participating in the Programme:

A. ICIE: Istituto Cooperativo per l'Innovazione, Rome, Italy. Also involved in the activity are researchers from the University of Rome; CIAM, an engineering company; TEP, a software company. ICIE is the coodinator of the Programme.
B. ABACUS: of the University of Strathclyde, Glasgow, United Kingdom. Also partecipating: Philips International BV, Eindhoven, Holland; Technische Hogeschool Eindhoven, Holland; University of Liege, Belgium.
C. CSTB: Centre Scientifique et Technique du Batiment, Nantes, France. Also partecipating ENTPE-LASH: Laboratoire des Sciences de l'Habitat de l'Ecole Nationale des Travaux Publics de l'Etat, Baulx-en-Velin, France.
D. SIV: Societa' Italiana Vetro, San Salvo, Italy. Also partecipating researchers from the University of Venice, Italy, Munchen, Fed, Rep. of Germany and Sevilla, Spain.
E. RWTH: Rheinisch - Westfalisch Technische Hochschule, Aachen, Fed. Rep. of Germany.
F. ETSAB: Escuela Tecnica Superior de Arquitectura de Barcelona, Spain.

4. Daylight data

The work in this area is carried out by the CSTB group:

The lack of daylight data measurements in meteorological stations requires the setting of models which allow to calculate luminous data (sky luminance distribution) from energetic data, such as irradiance, sunshine duration, nebulosity /2/.

The first step has been the classification of all skies into 5 types:
- overcast sky
- intermediate overcast sky
- intermediate mean sky
- intermediate blue sky
- blue sky.

The classification is based on the value of the nebulosity index, IN, defined as the ratio (1-CRM)/(1-CRT), where CRM and CRT are, respectively, the measured and the theoretical (clear sky) cloud ratio, that is, the ratio between the diffuse irradiance and the global irradiance on a horizontal surface, quantities that are measured in many meteorological stations.

The proposed classification is the following:

Overcast sky	IN : 00.0 - 00.5
Intermediate overcast sky	IN : 00.5 - 0.20
Intermediate mean sky	IN : 0.20 - 0.70
Intermediate blue sky	IN : 0.70 - 0.90
Blue sKY	IN : 0.90 - 1.00

The work under way in this area is to generate algorithms that allow to estimate sky luminance as a function of sky type, based on 15 months of measurements in Nantes, where global and diffuse horizontal irradiances have been measured every ten minutes, together with sky luminaces in 11 directions /3/.

Then, daylight data will be generated for all European locations, where irradiance data are available, using these algorithms.

At present, this has been done for the locations where hourly measured values of global and diffuse irradiances are available, that is for 12 of the 29 European stations of the

Test Reference Year.

5. Luminance models

The activity is carried out by ABACUS.

Objective of the work in this area is the selection of models and algorithms to estimate illuminance and irradiace on any surface, both internal and external, based on the data relative to horizontal surfaces generated as described in the previous chapter, and in the presence of configurations characterized by multiple reflections. These data are necessary for a joint simulation, for instance on a hourly basis, of the luminous and thermal aspects of a building's behaviour.

A critical review of existing literature has been started /4/.

6. Materials

Many new materials are being made available to daylighting designers. In order to allow the designer to quickly select the most appropriate solution to his problem, a set of parameters defining the specific role of each material has to be elaborated. At the same time, such parameters should provide all information necessary to numerically simulate the reflection and trasmission of the light reaching the material /5/.

Generally speaking, it is necessary to know the angular distribution (intensity) of the luminous flux reflected or transmitted by a material, as a function of the angle of incidence of the incoming light.

The activity in this area is, thus, related to several aspects:
- identification of new opaque and translucent materials
- definition of parameters to characterize their properties at two levels: descriptive, for designers, and detailed, for numerical analyses
- definition of laboratory measurements procedures
- identification of sources of data on materials and of laboratories where said measurements are, or could be, performed
- gathering and organisation of data to produce usable information.

The following classification of materials is proposed:
- Opaque surfaces.
Only three parameters are found necessary to describe most common surfaces:

RD = Diffuse reflection coefficient
RS = Specular reflection coefficient under normal incidence
DIS = Dispersion coefficient; three levels : 0 = no disper-
     sion, that is, specular; 1 = low dispersion; 2 = high
     dispersion.
- Translucent materials.
At least the following parameters are necessary:
TS = Specular transmission under normal incidence
TD = Trasmission of diffuse light
DIS = Dispersion coefficient (same as above).

    Particularly interesting appears an entirely new family
of materials, the optical switching materials, that can either
respond passively to varying environmental forces, or be
actively controlled according to environmental conditions or
changing building requirements /6/.

    There are three main categories of such materials /7/:
- Electrochromic, where a transparent medium changes color due
  to an ion-insertion reaction induced by an instantaneous
  applied electric field. An example of such materials is
  nickel oxide, which can be switched from transparent to
  bronze or dark grey, depending on impurities, with a typical
  change in visible transmittance from 0.70 to 0.15.
- Photochromic, that is, materials that change their optical
  properties with light intensity, due to a reversible change
  of a chemical species between two energy states with
  different absorption spectra. This change can be induced by
  electromagnetic radiation. There are many materials of
  this type, organic, inorganic, usually with traces of heavy
  metals or hallogens, and doped or metal oxides glasses.
- Thermochromic, which change their properties with temperature, They can be combined for a wide range of transition
  temperatures, but so far, have shown nonreversible reactions
  with long cyclic lifetime. The transmission factor can vary
  in a range of 1 to 3.
Other interesting materials beeing investigated are:
- Prismatic surfaces, where the control over sunlight penetra-
  tion is reached through an increased sensitivity of the
  transmission factor to the incidence angle, usually using
  glass or acrylic Fresnel Type lenses.
- Holographic films, that intercept incident sunlight and
  diffract it in another direction. At present, such films
  can only be produced in limited sizes and create rainbow
  effect, being sensitive to wavelength, but deserve attention.

7. Components

The work in this area is carried out by ETSAB and RWTH.

On one hand, a classification is beeing prepared, aiming at various objectives. First, it permits the organized analysis of different systems existing in real buildings and projects. At the same time, it can be used to establish a consistent methodology to test physical models. Furthermore, to check the performance laws, it will generate an evaluation system that will include coefficients of utilization for the studied elements and components.

Components are defined by the function they perform. A component can be seen as an assembly of elements.

Elements can be classified as /8/: divisions, fins, overhangs, lightshelves, baffles, jalousies, awings, reflectors, blinds, shutters, louvers, curtains.

Elements can be analysed according to their position, mobility, orientation and optical properties (transparency and reflectivity).

A first proposal for classifying daylighting components defines two main groups:
- Pass-through components: devices which divide two lighting environments; they are designed to allow and regulate the passage of light from one environment to another. They can be classified according to their position (facade or roof) and orientation (vertical, horizontal, tilted), as well as according to shape, size and location.
- Conduction components: devices that guide and distribute light towards the interior of the building. They can be divided into sub-groups: courtyards, galleries, light spaces and light pipes. Possible characterizations can be made according to their shape, slenderness, reflectivity (both as quantity and as quality).

A Data Base is also being developed.

Algorithms to approximately evaluate the performance of significant components as a function of design parameters are being selected.

On the other hand, a more detailed analysis of selected components is also under way. The first of such components being investigated are reflective components for side-lit spaces (lightshelves, louvers and prisms). Design guidelines for each component, concerning dimensions, position, materials, construction, etc., are being developed, with attention for local climatic conditions.

The investigation is being carried out by means of scale models testing under real sky conditions /9/.

So far, combinations of Lightshelves and window sills have been tested.

8. Controls

"A daylighted building, no matter how well designed saves energy only if the daylighting can effectively displace electric lighting usage.... For significant savings to be achieved (in lighting and cooling energy and in peak-demand utility costs) electric lighting control must be automatic. Automatic controls are made up of a combination of a sensing device, to measure daylight, and a controller that either switches or dims the electric lighting." /10/

Responsible for the activity is the ICIE/CIAM group.

The field of investigation concerns hardware devices and software procedures for /11/:

- acquisition of data on natural and artificial illumination, as well as on other environmental quantities
- calculation of comfort parameters and comparison with set standards and objectives
- implementation of local and general management policies, by means of regulators applied to artificial illumination and air-conditioning systems, as well as actuators applied to movable and controllable daylighting components.

The research is following two main lines:

- gathering of information on existing control hardware and software
- design of an integrated control system, based on luminous and thermal comfort.

In the first area, information has been collected on illuminance sensors, control systems and starters.

A very detailed survey of light measuring instrument has been performed by the RWTH within the activity related to defining suggested characteristics of data acquisition systems for scale models /12/.

9. Artificial lighting equipment

Objective of the research in this area is the collection of information on lamps, luminaires and switching/dimming devices, in order to assess their integrability with daylighting strategies.

Responsable is the ABACUS/Philips Group.

The first results /13/ concern the creation of a database on light intensity distribution of different lamps and luminaires. The data in each record consist of:

- lamp and luminaire type
- dimensional and geometric characteristics
- optical total for optimal temperature conditions
- the light intensity distribution for 19 angles for up to

four different planes.

10. Systems

The activity is carried out by SIV and RWTH.

Aim of the activity in this area is the identification, analysis and description of high quality buildings with significant daylighting features and contribution, to be shown as "examples" or, better, as "typological precedents", of good daylighting design /14/.

A survey form has been prepared, to collect all characteristics and features of selected buildings, as well as energy and comfort measured data, when available. Morphologically the building is described by grouping its characteristics into four typological levels:
- level 1: position of openings as seen from the interior; depth of room with respect to natural light sources; movables shading devices.
- level 2: ratio of transparent surfaces to opaque surfaces; relationship of such ratio with climatic conditions; size of openings; ratio of surface to volume; shape; openings distribution with respect to internal organization and function.
- level 3: orientation of openings with respect to sun position and urban space; fixed external shading devices.
- level 4: distance between building and adjacent buildings; dimensional proportions of streets, type of urban structure.

11. Numerical simulation and design tools

Numerical simulation of daylighting systems and design tools are, in principle, two partially overlapping, but separate, research areas. Numerical simulation tools (codes) can be used to aid designers, but also as a research tool, without immediate design applications. Architecture and presentation of codes can be quite different, according to the intended user. On the other hand, design tools is a much broader family than just simulation tools: it includes graphical instruments, nomographs, scale models, simplified methods, etc. The two categories, though, are treated together in this paper, because the research has not yet reached a point where the distinction is necessary.

The activity in this area, carried out by ABACUS, SIV, RWTH and ICIE/TEP, comprises several actions:
- critical review of existing computer codes and design tools
- detailed analysis of selected computer codes and design tools

- comparison of codes and design tools
- validation of computer codes, using data from scale models
  and actual buildings
- development of daylighting versions of thermal and energy
  analysis computer codes
- selection of suggested European design tools.

In the first area, information about many existing
computer codes is being collected and organized so as to make
it possible to compare and criticize them, and to find out
merits and deficiencies /15, 16/.

In the second area, the ESP code /17/ has been selected
as the thermal code to be implemented with daylighting
simulation capabilities. ESP appears particularly suited,
because, besides being able to simulate the thermal behaviour
of buildings with any geometry and of any type of plants,
already has provisions for simulating: spectral analysis of
glazing systems, time varying shading caused by site
obstructions, solar ray tracing, detailed view factors
assessment, distributed control systems.

The development of the daylighting expansion of ESP,
called DIM, has already started /18/.

Scale models are a very powerful research design tool, as
light studies, unlike many other studies, do not require
scaling correction: a model that exactly duplicates a full-
scale space, with the same materials, under similar sky conditions, provides identical results. Furthermore, most designers
are already familar with construction and use of scale models.

The activity in this area will provide guidelines for the
construction and instrumentation of physical models, both of
components and of buildings /9/.

12. Economics

The activity, carried out by ICIE/CIAM, is following two
main lines.

The first one concerns the identification of methods and
algorithms for evaluating the economic impact of daylighting
choices. Different cost-benefit analysis approaches can be
adopted, providing designers and investors with different
answers.

The second one is related to gathering information and
defining a procedure for defining "costs" and "benefits".
Thus, a first study on the rates structure of electric energy
in Italy and in France has been made /19/, as to provide the
basis for converting estimated energy savings in monetary
units. A study on costs of typical "daylighting" materials
and components in various European coutries will also be

conducted, as to provide data for estimating the incremental
costs possibly associated with an extensive use of daylighting
techniques.

References

1. Fanchiotti, A., "European Concerted Action Programme on
   Daylighting", Report presented at the contractors
   coordination meeting for Solar Energy, Research Areas A
   and D, Bruxelles, November 13-14, 1986.
2. Perraudeau, M. (CSTB) P., "Progress Report on Daylight
   Data and Algorithms", Nantes, France, September 1987.
3. Perraudeau, M. and Chauvel, P., "One Year's Measurements
   of Luminous Climate in Nantes", 2nd International Daylighting Conference, Long Beach, California, November
   1987.
4. de Wit, M. (ABACUS/FAGO), "Progress Report on Luminance
   Models and Algorithms", Barcelona, Spain, March 1987.
5. Fontoynont, M. and Paule, B. (CSTB/ENTPE), "Progress
   Report on Materials", Nantes, France, September 1987.
6. Johnson, R., Connell, D., Selkowitz, S. and Arasteh, D.,
   "Advanced Optical Materials for Daylighting in Office
   Buildings", Proceedings of the 10th National Passive Solar
   Conference, Raleigh, North Carolina, October 1985.
7. Selkowitz, S., Hunt, A., Lampert, C. and Rubin, M.,
   "Advanced Optical and Thermal Technologies for Aperture
   Control", Passive and Hybrid Solar Energy Update,
   Washington, D.C., September 1984.
8. ESTAB Group, "Progress Report on Components", Nantes,
   France, September 1987,
9. Willbold-Lohr, G. (RWTH), "Progress Report on Components
   and Scale Modeling", Nantes, France, September 1987.
10. Moore, F., Concepts and Practice of Architectural Day-
    lighting, Van Nostrand Reinhold Company, New York, 1985.
11. Monzani, A. and Becchi, E. (ICIE/CIAM), "Progress Report
    on Controls", Barcelona, Nantes, France, september 1987.
12. Willbold-Lohr, G. and Kielgas, K. "Survey of Light
    Measuring Instruments", Nantes, France, September 1987.
13. Tenner, A, (ABACUS/Philips), "Progress Report on Artifi-
    cal Lighting Equipment", Barcelona, Spain, March 1987.
14. Los, S. and Nasi, M., (SIV), "Progress Report on Systems
    and Design Tools", Barcelona, Spain, March 1987.
15. Matteoli, S. and Fanchiotti, A. (ICIE/TEP), "Progress
    Report on the Review of Existing Computer Codes", Nantes,
    France, September 1987. 16. ABACUS Group, "Progress Report
    on Numerical Simulation", Barcelona, Spain, March 1987.
17. Clarke, J.A., "A Technical Overview of the ESP System",

ABACUS CAD Unit, Glasgow, UK, March 1987.

18. Maver, T.W., Clarke, J.A., Stearn, D.D. and Kim, J.J., "Integrated Daylight and Artificial Lighting Model", Final Report for Grant GR/C/95541, ABACUS, University of Strathclyde, Glasgow, UK, 1987.

19. Bresciani, E. (ICIE), "Progress Report on Economic Methods and Data", Nantes, France, September 1987.

PROJECT MONITOR

Author   : D Turrent

Contract Number   : EN3S-0044-UK-(H)

Duration   : 36 months
                        1 April 1986 - 31st March 1989

Total Budget   : £ 279,700
                        CEC Contribution : £ 279,700

Head of Project   : D Turrent

Contractor   : The ECD Partnership

Address   : 11-15 Emerald Street
                        London
                        WC1N 3QL

Summary

Work on Project Monitor commenced on the 1st April 1986 and will continue for three years until March 1989. During this period a series of brochures will be produced on the major passive solar projects currently being monitored in Europe. These will include new build housing and refurbishment projects as well as some non-domestic building types such as schools and offices. The main aim is to produce a high quality brochure for each project which will be attractive to architects and therefore increase awareness of the benefits of passive solar design within the architectural profession. Active participants in the project include: Belgium, Denmark, France, Germany, Italy, Spain, Portugal, the Netherlands and the United Kingdom. The first seven brochures have been printed and will be distributed to 2,500 architects and engineers in the next few weeks.

1.1 Aims and Objectives

The main aim of Project Monitor is to provide architects and other building professionals with firm evidence of the benefits ofpassive solar design principles. In order to do this a number of built and monitored projects have been selected for publication in a series of colour brochures.

The overall design concept for the brochures was arrived at after a market research study which showed that architects tend to have a preference for visual images over words or numbers. Furthermore, their attention span is rather short and therefore technical documents must have a strong visual appeal in order to retain their interest.

1.2 Method of Work

The contract is being co-ordinated by The ECD Partnership, with material being supplied by sub-contractors in Belgium, Denmark, France, Germany, Italy, The Netherlands, Spain and Portugal. At earlier meetings of the group, a total of 50 projects were short-listed, based on the following criteria:

-   There should be a variety of building types

-   Housing schemes and retrofit projects are preferred to one off houses.

-   Projects should have some architectural merit.

-   Recent projects are preferred. Projects completed before 1982 will not be included.

-   The latest period for monitoring will be 1986/87 heating season.

-   The monitored data must be sufficient to calculate the solar contribution.

-   Good quality photographs and architectural drawings must be available.

The majority of the projects slected are in fact new housing schemes, although there are several examples of refurbishment using passive solar features and there are several office buildings, a sports building and a hospital included.

These projects have been organised in a publication sequence and ECD as co-ordinators have issued instructions to all sub-contractors specifying the quality and quantity of material required. For editorial work and the production of art-work, ECD employ a project manager and a graphic designer who liaise closely with the sub-contractors individually. ECD also liaise with University College Dublin on the distribution aspects and with the typesetters and printers.

1.3 Results Achieved

3,000 copies of the first seven issues have now been printed and these are being distributed to a mailing list provided by University College Dublin. To-date, this list contains 2,531 names and addresses including 1,730 architects. The breakdown is as follows:

Belgium	53
Denmark	35
France	59
Greece	260
Ireland	157
Italy	134
Luxembourg	2
Netherlands	264
Portugal	304
Spain	287
UK	444
West Germany	422
Outside EEC	110
Total	2,531

A Project Monitor binder is also being made available to receipients at a modest cost, so that eventually the whole set of brochures can be filed in a technical library.

1.4 Future Projects

Future projects for publication include the following:

Issue 8:   JEL Building, Stockport UK. A low energy industrial building with 36% office content and a total area of 2,000m2. The building has a south facing atrium which is used to pre-heat ventilation air to the industrial floorspace.

Issue 9:   Casa Termicamente, Oporto, Portugal. A
            two-storey detached house with a floor area
            of 135m2, built in heavyweight construction
            and employing a variety of heat storage
            systems.

Issue 10:   San Cugat, Barcelona, Spain. A short
            terrace of three-storey family houses
            incorporating double height sunspaces.

Issue 11:   Les Basses Fouassieres, Angers, France. A
            development of twenty seven, three and
            four-bedroom family houses, built in short
            terraces incorporating double height sun-
            spaces and trombe walls.

Issue 12:   Briarcliffe House, Farnborough, UK. An
            award-winning office building on four
            floors giving a total area of 10,000 m2.
            Its main feature is a double skin glazed
            wall which both helps to pre-heat air
            supply to the air conditioning system and
            acts as an acoustic barrier to external
            noise sources.

Issue 13:   Hoofddorp, The Netherlands. This
            development consists of six rows of eight
            terrace houses and one row of four duplex
            apartments. Each one is designed by a
            different architect and employs a variety
            of energy conservation, passive solar and
            active solar techniques.

Issue 14:   Marostica, Italy. The scheme consists of
            24 houses in three terraces and 16 flats in
            a four-storey block. all the dwellings are
            designed along passive solar principles
            with the houses featuring 'solar chimneys'
            and the flats having conservatory
            extensions to the living rooms.

Issue 15:   Lawrie Park Road, London, UK. This scheme
            consists of 18 two-storey family houses set
            out in three south facing terraces. The
            houses incorporate conservatory extensions
            with thermosyphon solar water heating
            panels mounted in the conservatory roofs.

Issue 16:   Lievre d'Or, France. This project involves
            the refurbishment of an estate of 593
            dwellings built in the 1960's. Solar
            techniques have been built into 211
            dwellings including: enlargement of south
            facing windows, addition of balcony -
            conservatories and conversion of gable
            walls into solar collectors.

Issue 17:   Baggensgade, Denmark. This is another
            refurbishment project, involving the
            addition of new glazed sunspaces on an
            existing five-storey residential building.

Issue 18:   Ecole Tournai, Belgium. A new school
            building designed around a central sunspace
            with smaller sunspaces around the building
            and built in heavyweight construction. An
            active solar air heating system and rock
            bed storge is also included.

Issue 19:   Arona, Italy. A new build block of 24
            apartments incorporating energy
            conservation measures, passive solar
            features and active solar systems for space
            and water heating.

Issue 20:   Jeux du Soleil, France. A new build
            housing scheme of 33 units incorporating
            high levels of thermal insulation, sun
            spaces, solar air collectors and heat
            pumps.

Issue 21:   Lundstuhl, West Germany. A scheme of five
            single-storey family houses, incorporating
            direct gain passive features, sunspaces and
            active solar water heating systems.

Issue 22:   Eissler Stradtsparkasse, West Germany.
            Also located in Landstuhl, this is a low
            cost housing scheme of eight units with
            direct gain, solar water heating and air
            heaging systems.

Issue 23:   Peterborough, UK. A short terrace of three
            new houses with conservatories and
            integrated solar air collectors on the main
            roof and first floor facade.

Issue 24:   Los Molinos, Spain. An experimental new
            build house employing a variety of passive
            solar techniques including a natural
            heating and cooling system, suitable for
            the mediterranean climate.

Issue 25:   Smakkebo, Denmark. A new build scheme of
            55 dwellings including high levels of
            insulation, sunspaces and thermosyphon
            solar water heating.

Issue 26:   Barbarcina, Pisa, Italy. A new apartment
            block of 45 flats with sunspaces and active
            solar space and water heating.

Issue 27:   Lou Souleu, France. A new apartment block
            of 22 flats designed around a central
            atrium.

Issue 28:   Haus Budel, Germany. A timber frame single
            family house with a sunspace and ground
            source heat pump.

MONITORING OF TWO PASSIVE SOLAR HOUSES

Contract   : EN3S-110-GR(110)
Duration   : 30 Months   1 September 1987 - 28 February 1990
Total Budjet   : 36.108 ECU   CEC Contribution : 24.108 ECU
Head of Project : Dr. E. Andreadaki Department of Architecture
                                     Faculty of Technologies
Contractor   : Aristotles University of Thessaloniki
Adress   : Aristotles University of Thessaloniki
                    Faculty of Technologies Department of Architecture
                    54006 Thessaloniki. Greece

Summary

The monitoring of passive solar buildings contributes to the techniques' development and provides experience to improve the passive solar systems. Several passive solar buildings, mainly private houses,
have been constructed in Greece, however monitoring hasn'nt been realized yet. A contract has been recently signed between the University
of Thessaloniki and the European Communities to execute the project
"Monitor of two Passives Solar Houses", one in the city of Athens and
the other in a suburb of Thessaloniki. The houses will be equipped
with an automatic data logger and a range of sensors to record temperatures, solar radiation, wind direction, speed etc. It has been decided to carry out the monitoring in different climatic conditions,
latitudes and building regulations in order to obtain different results, which will be compared. This project will be allied with the
"Project Monitor" programme which has been contracted between the ECD
Partnership and the CEC. I believe that the existing thermal methods
which evaluate the thermal behaviour of passive solar buildings
should be adapted to the greek climatic conditions in order to satisfy both the required heating in winter and cooling in summer.

- 1.1. Introduction

The monitoring of passive solar buildings contributes to the development of techniques' and provides experience to the architects about the improvement of passive solar systems.

The passive solar design of buildings becomes more and more acceptable in Greece, namely architects, builders and the population. There are many reasons for this acceptance, I will only refer the following:

- considerable energy conservation.
- buildings'adaptation to the environment, the local climate and sometimes the regional architecture.
- contribution to the clarification of the atmosphere.

This presentation refers to a research programme for the "Monitoring of two Passive Solar Houses" in Greece. The contract is recently signed (in September) between the University of Thessaloniki and the European Communities. The programme is related to the "Project Monitor" entrusted to the ECD Partnership by the Commission.

We hope that the present project will be included in the "Project Monitor" programme in the future.

- 1.2. Description of the two Passive Solar Houses

The monitoring will take place in two private houses.

a. A four storied house in the city of Athens, Ampelokipi, with total area of 210 sq. metres and a volume of 900 c. metres.

The architect-designer of the house, Katerina Spyropoulou, declares that she had to solve several problems such as orientation, safety requirements and economic restrictions.

- Site plan: the building lot is extremely small, only 99 sq. metres, it is situated in a densely built area in the city of Athens. Many restrictions arised from the urban and building regulations (fig. 1).

- House plans: the ground floor is used for central heating installation and other supplementary facilities because it cannot receive any solar gain due to the height of the neigbouring buildings. The main house is raised from a height of 5.5 metres above street level. It consists of three floors with an internal courtyard (fig. 2) orientated to the south in order to let direct sunlight penetrate to the north side of the house in winter.

Both the main south facade and the south courtyard are used for passive solar energy collecting purposes (fig. 3,4).

- Passive solar systems: a sunspace stretches from a height of 9 metres (mezzanine floor) up to the two bedroom floors. The total sunspace area is 18 sq. metres and its upper part is tilted by an angle of 20° degrees, to increase sun exposure. Warm air flows directly to the bedroom floors, while two small fans force the circulation of hot air to the living room area (fig. 3). Two surfaces situated in the courtyard are used as Trombe walls (fig. 2,4). There are dampers to control air flow in each room. Openings are situated opposite to the trombe walls giving the opportunity of ventilation during summer (fig. 5). In addition rollover shutters exist between windows and the glazing of the trombe walls to prevent heat loss or over-heating.

- Energy calculations: calculations had to take into account the effect of shadows cast by the building itself on the collecting surfaces. The total solar energy collected by the three passive systems covers about 60% of the required energy for space heating (fig. 6). The rest will be provided by the central gas heating system and the fireplace.

The project started in 1979-80. The house is inhabited since last October (1986). The tenants declare that the sunspace has rendered more heat than the trombe walls, because rollover shutters hadn't been put. The temperature on the upper floors was about 23°C, while the living room area in the first floor had a reduced temperature of about 20°C or even below this. The central heating system has been in operation for 3 hours, between 5-8 a.m., in the evening and one hour in the morning. Note that last winter was very cold for the usual climatic conditions in Greece.

The house will be equipped with an automatic data logger 32 KB memory, 4 signal transducer for PT100 and global radiation. The proposed registration technic is a hardware memory. The memory card will be exchanged every week based on 15 min. averages of the recorded values.

The sensors which will be installed are:

- solar radiation (global). - ambient air temperature.
- room temeperature, 10 units. - sunspace temperature, 2 units.
- trombe wall temperature, 2 units. - wind speed and direction.
- 4 status of windows. - electricity meter.
- gas heat meter. - mounting material.

b. <u>A private house in a suburb of Thessaloniki (Thermi)</u>.

The total area of the house is 200 sq. metres and its volume about 800 c. metres. Mrs. Axarli and Mr. Antoniou are the architects-designers of the building.

- Site plan: the house is situated in a free area without any urban restrictions. It has been easier for the house to be orientated to the south. The direction of the long axis is going from the east to the west, such as the southern elevation has a length of 18 metres and all passive solar systems are incorporated on it.

- House plans: the ground floor consists of the living room areas, the dining room, the kitchen and a guest's room (fig. 7). In the first floor there are the bedrooms and the other facilities (fig. 8). The central heating system is installed below the ground floor.

- Passive solar systems: a sunspace attached to the living room with a total area of 18 sq. metres. The hot air flows into the living space through the windows and part of it reaches a bedroom on the first floor (fig. 9,10). There are five (5) trombe walls, three (3) in the ground floor and two (2) in the upper floor with a total area of about 33 sq. metres (fig. 7,8). The northern elevation has a restricted number of openings.

- Energy calculations: the total energy collected from the passive solar systems covers about 55% of the required energy for space heating. The rest will be provided by the central oil heating system. The total house is built with concrete and massive brick, so a great accumulative mass absorbs solar energy heat during winter days.

The house hasn't been inhabited yet.

This house will also be equipped with an automatic data logger and a range of sensors as the previous house.

The monitoring will take place first at the house in Athens for a year and a half and then the total equipment (mainly the data logger which is expensive) will remove at the house in a suburb of Thessaloniki.

- 2. <u>Conclusions</u>

It is obvious that we haven't any results from the monitoring (because the contract was recently signed). The only indications, refferring to initial goals are the following:

- the expected results of the monitoring will verify or not the calculations made with the simplified methods of the buildings'thermal performance and find out the degree of thermal comfort into the passive solar buildings.
- this evaluation contributes to improve passive solar systems, especially the quantity and distribution of the accumulative thermal mass. These results are very important for the greek climatic zones where the half of the year is hot and the problem of the natural cooling is of the same importance as the problem of the solar heating.

HOUSE IN ATHENS

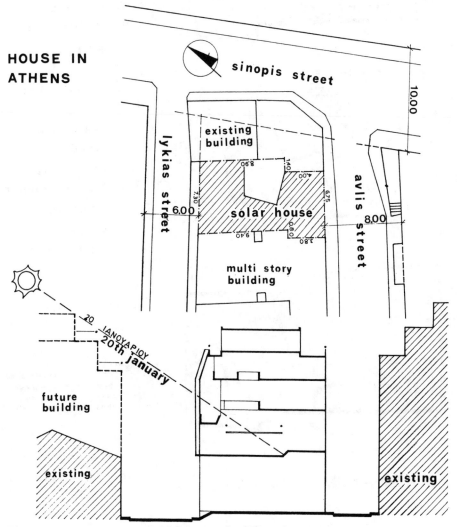

Fig. 1. The site plan of the house in Athens.

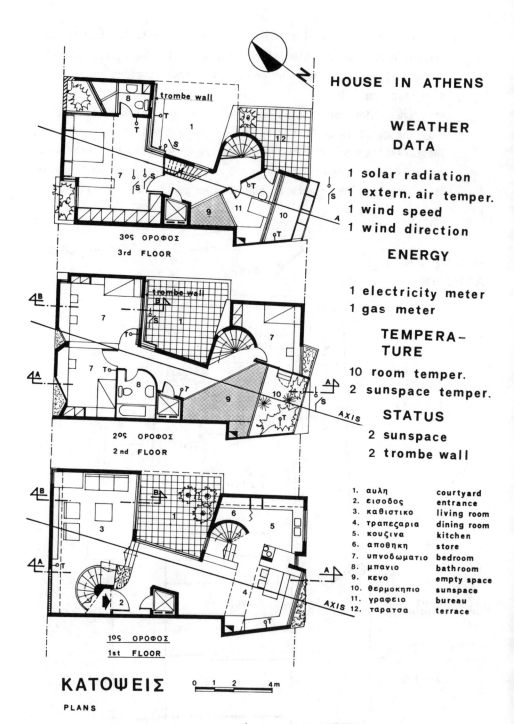

HOUSE IN ATHENS

WEATHER DATA

1 solar radiation
1 extern. air temper.
1 wind speed
1 wind direction

ENERGY

1 electricity meter
1 gas meter

TEMPERATURE

10 room temper.
2 sunspace temper.

STATUS

2 sunspace
2 trombe wall

1. αυλη	courtyard
2. εισοδος	entrance
3. καθιστικο	living room
4. τραπεζαρια	dining room
5. κουζινα	kitchen
6. αποθηκη	store
7. υπνοδωματιο	bedroom
8. μπανιο	bathroom
9. κενο	empty space
10. θερμοκηπιο	sunspace
11. γραφειο	bureau
12. ταρατσα	terrace

3ος ΟΡΟΦΟΣ
3rd FLOOR

2ος ΟΡΟΦΟΣ
2nd FLOOR

1ος ΟΡΟΦΟΣ
1st FLOOR

ΚΑΤΟΨΕΙΣ

PLANS

Fig. 2. House plans, 1st floor, 2nd floor, 3rd floor.

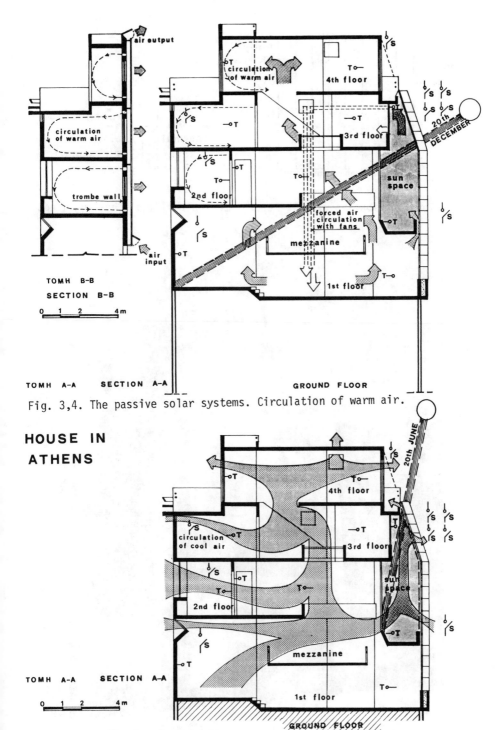

air output

circulation
of warm air

trombe wall

air input

TOMH B-B
SECTION B-B

0 1 2   4m

○T circulation
of warm air

4th floor

S

○T

3rd floor

sun
space

2nd floor

forced air
circulation
with fans

mezzanine

1st floor

TOMH A-A   SECTION A-A

GROUND FLOOR

20th DECEMBER

Fig. 3,4. The passive solar systems. Circulation of warm air.

HOUSE IN ATHENS

4th floor

circulation
of cool air

3rd floor

sun
space

2nd floor

mezzanine

TOMH A-A   SECTION A-A

0 1 2   4m

1st floor

GROUND FLOOR

20th JUNE

Fig. 5. Ventilation during summer.

– 139 –

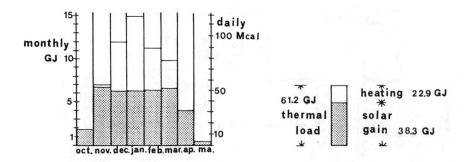

Fig. 6. The contribution of solar energy to the building's heating needs.

External view of the house.

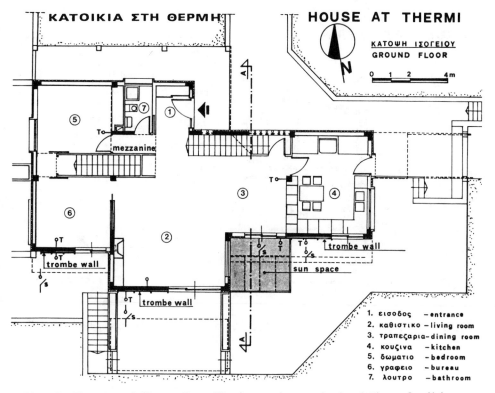

ΚΑΤΟΙΚΙΑ ΣΤΗ ΘΕΡΜΗ

HOUSE AT THERMI

ΚΑΤΟΨΗ ΙΣΟΓΕΙΟΥ
GROUND FLOOR

0   1   2   4m

N

mezzanine

trombe wall

trombe wall

trombe wall

sun space

1. εισοδος   — entrance
2. καθιστικο — living room
3. τραπεζαρια — dining room
4. κουζινα   — kitchen
5. δωματιο   — bedroom
6. γραφειο   — bureau
7. λουτρο   — bathroom

Fig. 7. The ground floor plan. The house in a suburb of Thessaloniki.

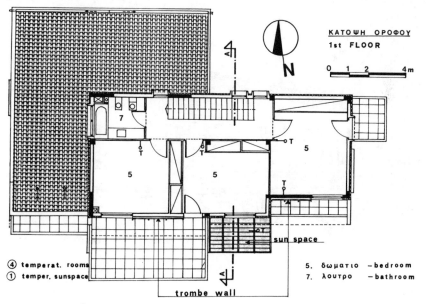

ΚΑΤΟΨΗ ΟΡΟΦΟΥ
1st FLOOR

0   1   2   4m

N

sun space

④ temperat. rooms
① temper. sunspace

trombe wall

5. δωματιο — bedroom
7. λουτρο   — bathroom

Fig. 8. The 1st floor plan.

HOUSE AT THERMI

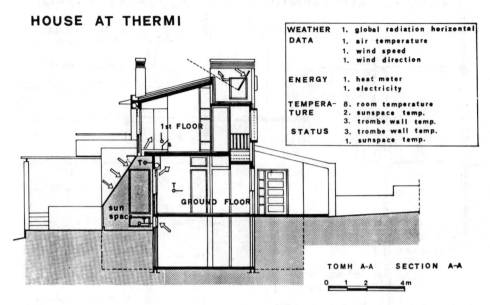

WEATHER	1.	global radiation horizontal
DATA	1.	air temperature
	1.	wind speed
	1.	wind direction
ENERGY	1.	heat meter
	1.	electricity
TEMPERA-	8.	room temperature
TURE	2.	sunspace temp.
	3.	trombe wall temp.
STATUS	3.	trombe wall temp.
	1.	sunspace temp.

1st FLOOR

GROUND FLOOR

sun space

TOMH A-A   SECTION A-A

0   1   2   4m

Fig. 9. The section of the house. Hot air input from the sun space.

HOUSE AT THERMI (THESSALONIKI - GREECE)

Fig. 10. The external view of the house.

TECHNOLOGY TRANSFER FROM THE SOLAR R & D COMMUNITY TO THE EUROPEAN BUILDING PROFESSIONS: SOLINFO

Contract Number: EN3S-0087-IRL

Duration: 36 months   01.08.1986 - 31.07.1989

Total Budget: 530,000 ECU   CEC Contribution: 100%

Head of Project: J. Owen Lewis

Project Manager: Shane O'Toole

Contractor: University College Dublin

Address: Energy Research Group
                        School of Architecture
                        University College Dublin
                        Richview   Clonskeagh
                        Dublin 14   Ireland

Summary

The CEC R & D sub-programme, Solar Energy Applications in
Buildings, faces an unusually difficult task in the dissemination of
its results because of the diversity and size of its target audience.
SOLINFO aims to promote the transfer of information from the solar
research community to the building design and construction
professions.   It concentrates on the areas of improving our
understanding of information requirements, supporting
undergraduate education and mid-career education, providing aids
to professional practice, and speeding-up research dissemination.
A series of national review tasks has been completed and an
analysis of the findings prepared.   An expert team of national sub-
contractors is being appointed both to form the basis of an
information network and to act as an advisory group for the project
as a whole.   There is also a series of related tasks.   Priority areas
of work include: identification of information content and format
required by users;   establishing information networks between
'centres of excellence';   compiling a catalogue of selected passive
solar components;   and implementing a series of CEC research
digests.   Close linkages exist between the SOLINFO and ARCHISOL
projects.

1.0 PROJECT DESCRIPTION

1.1. Introduction and Aims

A very great amount of information is being generated in research and development on solar energy applications in buildings.   Yet this knowledge is not being implemented in the industry quickly or extensively enough.   This project is designed to promote the transfer of information from the solar research community to the building design and construction professions.

The overall objective is to improve the understanding of the particular information requirements of the different building designers and constructors (whether planners, urban designers, architects, building engineers, builders and developers) at the various stages of the building process, and to make available this information in forms convenient for use by practitioners and students.

1.2 Methods

The project tasks as set out in the research contract have been subject to some minor revisions resulting from comments received at project meetings, and some updating.   The work envisaged falls into five areas:

* Information Requirements
* Undergraduate Education
* Mid-Career Education
* Professional Practice
* Research Dissemination.

Anticipated results of the work are:

* A report on energy-related information needs of the building process and the building professions, and continuing review of the project;
* Information on innovatory undergraduate teaching of low-energy design and solar heating and daylighting, and generation of teaching materials (including printed and audio-visual media);
* Information on innovatory mid-career teaching, and promotion of handbook and workshops;
* A revised European Passive Solar Handbook, together with design and specification information;
* CEC solar research digests and more rapid dissemination of research results.

The scope and content of the tasks was detailed at a previous Contractors' Meeting (1).

2.0 PROGRESS TO DATE

2.1 Information Requirements

Comparatively few studies of either energy-related education and training needs of architects, engineers, builders and the trades, or information requirements of the building design process related to energy have yet been undertaken, although, as part of an overall trend (2), some improvements can be seen in recent times.

Recognising the importance of building on, rather than duplicating, existing relevant work in this area, it was decided that a preliminary national review task should be carried out in each Member State. The intention was that analysis of the review task reports would assist in identifying institutions and individuals with whom to collaborate.

The authors of the national review tasks (see Appendix I) were requested, inter alia, to:

- Identify previous studies of energy-related education and training needs of architects, engineers, builders and the trades;
- Identify previous studies of information requirements of the building design process related to energy;
- Identify previous market studies of 'energy and building' products and services;
- Summarise existing programmes in the formal teaching of low-energy design and solar heating technology;
- Summarise existing programmes and experience in mid-career training in low-energy design and solar heating technology;
- Collect examples of nationally-produced passive solar design aids (i.e. professional handbooks/manuals and microcomputer software).

Certain of the information thus obtained would also be of value to ARCHISOL (3). The reports of the national review tasks were presented in their preliminary versions at an initial project meeting in Dublin on 13 and 14 October 1986. Most of the reports were subsequently added to, and furnished in their final form in early 1987. The reports were then analysed (4) in order to clarify the most effective forms of intervention and support, and to provide a framework for the remainder of this project and for ARCHISOL. A procedure for continuously up-dating the information obtained through the national review tasks is being implemented. There was an informal project meeting in Munich on 9 April 1987 during the European Conference on Architecture. The next project meeting is planned for Porto on 26 and 27 November 1987.

An important task now underway aims to develop a 'specification' for the format of building design aids, aimed at facilitating the better use by architects, in particular, of information on solar energy utilisation. It is widely accepted that 'information transfer' from the scientific community to practising architects is problematical. Analysis of this fact reveals two contributory factors:

- The inappropriate form of technical information in relation to the design process;

• The 'learning styles' of building designers make special demands on the ordering and presentation of information.

It follows that the 'specification' will be a prescription for the form of information to be disseminated, based on an evaluation of existing formats and of existing research into the use by architects of technical information.

While it is already clear that no single format will meet the range of demands arising from different stages in the design process, and from variations in 'modus operandi' of design professionals, there are distinct conveyences of experience evident in research or information use.

Studies of information use by architects indicate that architects rarely use research publications and that, for materials to contribute to effective information transfer, their presentational style must encompass brevity and clarity of text, good visual information and have a familiar layout style. The context must allow new material to be related to known ideas and contexts, and materials must enable designers to find appropriate answers quickly and without undue searching. In addition, it is clear that architects draw on different types of information source, and that the learning that takes place tends to be private. Technical information must be complete in the sense of allowing dependable specification. Generally, information is valued which allows architects to develop a 'feel' for the task facing them. 'Design standing' is important, and architects tend to value a few key individuals or organisations.

At the same time, there is developing insight into the relationship between technical information and the design process, and of the type of problem which needs to be confronted. In particular, it has been asserted that many applications of rules of thumb will be necessary before intuitive use of energy-related concepts can co-exist with considerations of space and form. In this respect it has been suggested that there is a need for graphic tools which relate quantitative information about performance to traditional concerns such as scale and proportion.

There is growing insight into the 'learning styles' of designers, and the proposition has been advanced that material needs to be specifically tailored to a matrix of learning styles. This in turn leads to consideration of 'information styles' and also to the type of learning exercise conducive to assimilation of concepts and the development of ability to apply them in design.

The sub-task (5) aims to complete its work in January 1988. It is envisaged that the specification will be in the form of written guidelines, with existing material utilised as exemplifying the approaches put forward.

2.2 Undergraduate Education

Notwithstanding the successful conclusion of negotiations leading to the adoption in 1985 of a Directive on Architecture (6), the national review task reports demonstrate that wide differences exist across the Community regarding the manner and conditions in which architects are educated, particularly at undergraduate level.   In considering architectural education, frequently one is in a position where it is not possible to compare like with like, as between systems of architectural education, or even as between schools of architecture in the same country.   Pending the publication of a vade mecum of architectural education within the Community, it is likely to remain difficult to acquire anything other than the most general of appreciations regarding the diversity in, and the nature of, the formal teaching of architecture which currently exists.

Nevertheless, we may already say that, in general, low-energy design and solar heating technology do not form a specific individual element in the various undergraduate syllabi of European schools of architecture, but often exist within the framework of a larger lecture subject area.   Even where such education is available, it tends often to be a reflection of the enthusiasm and interest of individual teachers. Not surprisingly, there is little consistency in the courseware employed. Whilst a number of exemplary courses and individual teachers have been identified, it could be said that passive solar design only occasionally features as a distinct topic in studio design projects.   The notable exception to this general rule appears to be when a school takes the brief of one of the national or international competitions as the basis for a project.   In such cases, energy questions assume greater significance.

It is clear, therefore, that much can be done to aid improvements in the formal teaching of low-energy design and solar heating technology in European schools of architecture.   Perhaps, however, the Danish experience indicates the true extent of the educational task to be confronted: it certainly suggests there is a case for attempting to influence and alter widely-held preceptions about these topics.   All Danish schools of architecture offer courses or modules in low-energy design and passive solar heating, but this does not mean that all students take these subjects, because the Danish formal education system for architects (and engineers) is characterised by the fact that the individual students themselves select their courses or modules from a catalogue offered by the schools.   Their selections must, naturally, respond to a 'grid' which ensures the students experience a range of subject areas and disciplines.   To this end, the possibility for guidance is available during the selection process.   Why, then, does only a minority of students specialise in low-energy design and passive solar heating ? There are, undoubtedly, many reasons, but among them are the students' sensitivity and responsiveness to trends and fashions in architecture generally and, perhaps more particularly, the students' perception of the value and use of the particular courses or modules in their coming professional careers. So long as low-energy design and passive solar heating do not form part of the mainstream of architecture and building, it appears that they will tend to remain marginal in such an education system. This topic has been addressed in a paper (7) to the 1987 ISES Congress in Hamburg.

Analysis of the national review task reports has led to the identification of selected 'centres of excellence' among existing educational facilities across the community. A small workshop of distinguished architect/educators will take place in Paris on 5 and 6 October 1987. The workshop will promote contact between these 'centres of excellence' and provide an occasion for exchanging information on their experience and teaching materials. One of the aims of the workshop is to form a sub-group (drawn from individuals who are making substantial and innovative contributions to the preparation of resource material for teachers and instructors, and to the development of pedagogic approaches in mid-career eduation) to collate, examine and disseminate existing approaches of merit and thereby establish information networks which will continue the process of information transfer and education after the completion of this contract. The workshop should result in the identification of appropriate educational institutions which could be invited to develop resource material for teachers and instructors.

2.3 Mid-Career Education

Particularly interesting developments in post-graduate courses and how they relate to mid-career training seem to have taken place in the Federal Republic of Germany, where, in recent years, the declining volume of construction work has created stiff competition among architects, resulting, among other things, in greater specialistion in areas, for example, such as energy consultancy and energy management. Arising out of a research project (8) conducted in 1982 and 1983, a successful post-graduate one-year consultancy course has been developed by the Technical University of Berlin in such a way as to allow professionals from the construction industry to follow the course with a minimum of disruption to their ongoing employment. Most of the learning can be done at home, supplemented by a few days of intensive instruction at the University. A pilot project with a similar underlying concept has been operated for several years at the Gesamthochschule Kassel.

However, practical experiments within the Community in the area of information dissemination to the architectural profession seem, at this stage, to be most advanced in Denmark. Several years ago it became evident in that country that much of the building construction-related information generated by research and development activities was not being satisfactorily implemented and that a gap existed between the research and development communities and the design and construction professions.   A manual (9) was produced describing a method for the systematic dissemination of information relating to the results of research and development activities concerning the reduction of energy consumption for heating.   The manual describes how to plan information, evaluate results and their significance, and define target groups.   It also includes a complete catalogue of relevant Danish professional and lay organisations and media, with contact information, circulation figures and so on.   A development of this project, specifically aimed at passive solar heating, is being undertaken by a passive solar working group representative of both the research institutes and the architectural profession.   The first phase of this information plan resulted in the publication of a widely distributed booklet (10) containing a simple message, and was followed up by a media campaign and a series of information meetings.   The next phase of the information plan, now underway, involves the publication of a more specific booklet and follow-up courses for practitioners.

Whilst the numbers of activities in the area of mid-career training and the numbers of architects attending them, appear to be decreasing, at least in northern Europe, as the emphasis shifts away from design towards energy efficiency and manaegment, a different picture emerges for southern Europe: in Greece, for example, there continues to be a professional demand for seminars of up to one hundred and fifty hours in duration on passive solar design and bioclimatic architecture.   The enthusiasm and support of the professions in southern Europe (notably including new Member States Spain and Portugal) for mid-career training programmes in solar design reflects an existing widespread market-led demand for professional expertise in this area.

The first of the regional workshops to introduce the preliminary edition of the European Passive Solar Handbook is planned for Porto, early in 1988, to coincide with a locally-organised architectural competition.   The forthcoming workshop in Paris, already mentioned, will also address the question of mid-career education.

2.4 Professional Practice

The preliminary edition of the European Passive Solar Handbook was made available to participants at the European Conference on Architecture, held in Munich in April 1987.   There has been a sustained, if modest, level of demand for the Handbook in the interim. Due to the limited number of copies available, however, the Handbook is not being made available generally to individuals.   A dissemination plan has been prepared and is resulting in copies of the Handbook being placed in the libraries of schools of architecture across the Community -

concentrating initially on the member schools of the European Association for Architectural Education (EAAE) - where it will be a major resource to students of architecture participating in the Third CEC Passive Solar Competition. The remainder of the Handbooks are being retained for use in a series of regional workshops and for distribution to selected key audiences and professional libraries.

A sub-contract has been concluded with the ECD Partnership, London, for the production of a passive solar components catalogue. If the number of buildings using passive solar design is to be increased, designers must become aware of and familiar with design techniques, examples of good practice and the required technical equipment. Design tools and examples of good practice are being promoted by other actions within the current CEC Solar R & D programme. This sub-contract is intended to acquaint European architects with passive solar components. Project MONITOR will be the main source of information on products. The cataloque is to be suitable for use as a drawing board aid and will be approximately 120 pages in length. It will include examples of all major types of components required for passive solar buildings and will include as much information on experience in use as possible. The work is scheduled for completion at the end of October 1988.

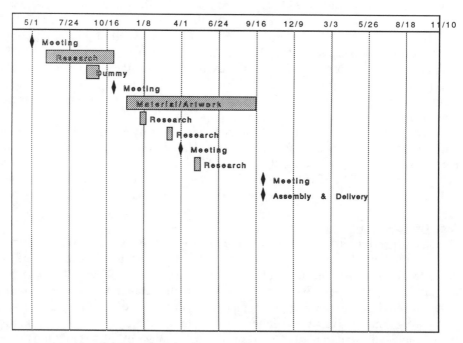

Table I: Passive Solar Components Catalogue: May'87 - Oct'88

2.5 Research Dissemination

A common format for research newsletters has been designed and distributed to all co-ordinators for use within the sub-programmes, Solar Energy Applications in Buildings and Solar Radiation Data.   Mailing lists for distribution of the various newsletters have been established and the Energy Research Group at UCD can provide a service to co-ordinators by providing pre-addressed labels as required.

A colour brochure describing the Solar Energy Applications in Buildings sub-programme has been prepared and was distributed at the ISES Congress (Hamburg) and at InterArch 87, the fourth World Biennale of Architecture (Sofia), both in September 1987.

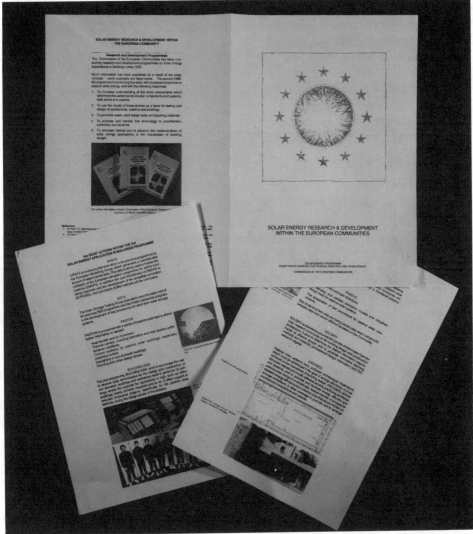

Fig 1: 'Solar Energy Applications in Buildings' Brochure

The next task is the preparation of a series of research digests for translation by, and circulation in, selected professional periodicals by agreement.   To this end, discussions with key journals and individuals have been initiated.   Publication of the first digest will follow the completion of the first series of newsletters.

A series of mailing lists for use by the Solar Energy Applications in Buildings sub-programme has been created.   The lists include selected professional architectural periodicals and specialist energy magazines, architects professional organisations and so on.   However, the largest of the mailing lists has been prepared to assist in the distribution of a series of brochures being produced by Project MONITOR.   It has been created in association with SOLINFO's network of national sub-contractors and through the active assistance of the Presidents of most national architectural organisations within the Community.   Methods have included direct mailing of 20,000 reply postcards, distribution of promotional information at architects' conferences and publication of news releases in a large number of professional periodicals.   The original aim was to create a mailing list of 2,000 European architects interested in passive solar design.   Already that number has been well exceeded.   The list contained 100 names at the end of 1986, 1,200 by May 1987 and achieved the target figure early in September 1987 (see Table II).   Of course, this list will have many further applications, not least in relation to the Third Passive Solar Competition and the 1989 CEC Conference on Architecture, to be held in Paris.

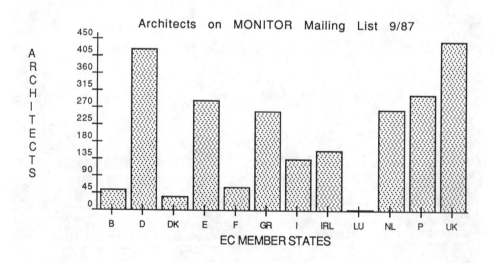

Table II

3.0 CONCLUSIONS

3.1 Present Status and Future Work

Technology transfer, which in the case of passive solar, is of particular importance (there are, for example, over 200,000 architects in Europe), has received little attention heretofore in Europe. However, the necessary objectives, methods and organisation structures for such a project have now been identified. An expert Committee has been established and we are now implementing the priority tasks of ;

- Identifying the content and format of information required by users;
- Modelling information networks of 'centres of excellence';
- Preparing teaching materials and design aids;
- Organising architects' workshops;
- Improving research dissemination.

4.0 REFERENCES

(1) Lewis, J. O., S. O'Toole. Technology Transfer from the Solar R & D Community to the European Building Professions: SOLINFO in Solar Energy Applications to Buildings and Solar Radiation Data - Proceedings of the EC Contractors' Meeting held in Brussels, Belgium, 13 and 14 November 1986. D. Reidel Publishing Company, Dordrecht, 1987

(2) Commission of the European Communities. European Symposium on the Utilisation of the Results of Public Research and Development, Luxembourg, 1986.

(3) Lewis, J. O., S. O'Toole. Architecture and Solar Energy: ARCHISOL in Solar Energy Applications to Buildings and Solar Radiation Data - Proceedings of the EC Contractors' Meeting held in Brussels, Belgium, 13 and 14 November 1986. D. Reidel Publishing Company, Dordrecht, 1987.

(4) O'Toole. S. Analysis of SOLINFO and ARCHISOL National Review Tasks. Energy Research Group, University College, Dublin, 1987.

(5) Kealy, L., architect and educationalist (Dublin) is conducting the sub-task in association with a consulting group comprising A. de Herde (Louvain-la-Neuve), P. Lodberg (Aarhus) and J. P. Traisnel (Paris).

(6) The Council of the European Communities' Directive of 10 June 1985 on the Mutual Recognition of Diplomas, Certificates and Other Evidence of formal qualifications in architecture, including measures to facilitate

the effective exercise of the right of establishment and freedom to provide services (85/384/EEC).

(7) Lewis, J. O., S. O'Toole and T. C. Steemers Education in European Solar Energy Utilisation. Hamburg, 1987.

(8) Entwicklung und Erprobung eines wissenschaftlichen Weiterbildungsprogramms mit dem Ausbildungsziel Energieberater. (Development and Testing of a Scientific Postgraduate Education Programme with the Objective of Training Energy Consultants).

(9) Steensen, P., M. Bendtsen, J. Knoblauch and P. Bindesbøll. Model for Informationsformidling indenfor området energiforbrug i bygninger (Model for Dissemination of Information Within the Area of Energy Consumption in Buildings). The Energy Research Programme of the Danish Ministry of Energy. Technological Institute and Department of Energy Technology, 1984.

(10) Steensen, P., E. Petersen, O. Jørgensen, P. Lodberg and B. Lundgaard. Passiv Solvarme: Hvorfor, Hvordan, Hvornår (Passive Solar Heating: Why, How, When). Teknologisk Institut, Danmarks Tekniske Højskole, Architektskolen i Aarhus and Kunstakademiets Arkitektskole, 1984.

APPENDIX 1: AUTHORS OF NATIONAL REVIEW TASKS

- BELGIUM AND LUXEMBOURG
  André DE HERDE and Anne MINNE, Architecture et Climat, Centre de Recherches en Architecture, Université Catholique de Louvain.

- DENMARK
  Poul LODBERG, Architekt, MAA, Aarhus.

- FRANCE
  J.P. TRAISNEL, AFME, Paris.

- FEDERAL REPUBLIC OF GERMANY
  Alex LOHR, Büro für energiegerechtes Bauen, Köln.

- GREECE
  A. N. TOMBAZIS, C. STAMBOLIS and E. ATHANASAKOS, Alexandros N. Tombazis & Associates, Athens.

- IRELAND
  Shane O'TOOLE, Energy Research Group, School of Architecture, University College, Dublin.

- ITALY
  Paola ARNEODO, Dipartimento de Energetica, Politecnico di Torino.

  Additional information regarding courses in architecture by Paola FRAGNITO, Progettazione e Ricerca, Milano.

- THE NETHERLANDS
  Joep HABETS, EGM Architecten, Dordrecht.

- PORTUGAL
  E. DE OLIVEIRA FERNANDES and Eduardo A. MALDONADO, Departamento de Engenharia Mecânica, Universidade do Porto.

- SPAIN
  Antonio PEREZ MORENO, Instituto de Energias Renovables, CIEMAT, Madrid.

- UNITED KINGDOM
  Nicholas BAKER and Dean HAWKES, Martin Centre for Architectural and Urban Studies, Department of Architecture, University of Cambridge.

ARCHITECTURE AND SOLAR ENERGY: ARCHISOL

Contract Number: EN3S-0088-IRL

Duration: 36 months   01.08.1986 - 31.07.1989

Total Budget: 580,000 ECU   CEC Contribution: 100%

Head of Project: J. Owen Lewis

Project Manager: Shane O'Toole

Contractor: University College Dublin

Address: Energy Research Group
                     School of Architecture
                     University College Dublin
                     Richview   Clonskeagh
                     Dublin 14   Ireland

Summary

ARCHISOL has the goal of advancing the introduction of energy-conscious building into the mainstream of architectural design. It seeks to introduce both practising architects and students of architecture to energy-efficient building and passive solar design in a stylistically neutral manner, to facilitate the construction of passive solar buildings and to establish energy efficiency as a criterion of architectural criticism. Methods include competitions, awards and exhibitions, and co-operation with European architectural periodicals. A series of national review tasks has been completed and an analysis of the findings prepared. An expert team of national sub-contractors is being appointed both to form the basis of an information network and to act as an advisory group for the project as a whole. There is also a series of related tasks. Priority areas of work include: extending contacts with journalists and critics; compiling an exhibition and catalogue of European regional vernacular energy-conscious buildings; developing competition calculation procedures and performance indices, and launching the Third CEC Passive Solar Design Competition; and initiating a series of regional competitions emphasising energy criteria. Close linkages exist between the ARCHISOL and SOLINFO projects.

1.0 PROJECT DESCRIPTION

1.1 Introduction and Aims

The elements of passive solar design cannot be considered only in their technical dimensions, as of their nature, passive systems have profound architectural implications. While architects can have a major influence on the efficiency with which energy is used in the servicing of buildings, technological feasibility does not in itself ensure the wide-spread implementation of new energy systems. This project has the goal of advancing the introduction of energy-conscious building into the mainstream of architectural design.

The overall objective is to modify attitudes and stimulate interest, to introduce both practicising architects and students of architecture to energy-efficient building and passive solar design in a stylistically neutral manner, to facilitate the construction of passive solar buildings, and to establish energy efficiency as a criterion of architectural criticism. The project also addresses the owners, managers and users of buildings.

1.2 Methods

The project tasks as set out in the research contract have been subject to some minor revisions resulting from comments received at project meetings and some updating. The work envisaged falls into five areas:

- Consultation
- Professional Media
- Lay Literature
- Competitions
- Realisation.

Anticipated results of the work are:

- A continuing review of the project and its relevance; and debate on regional energy-efficient buildings;
- Extensive coverage of energy efficiency and energy-efficient building in European professional media; and a wider awareness of the rich heritage of traditional exemplars;
- Improved awareness of solar heating and daylighting among potential clients and developers;
- Increased knowledge and awareness of passive solar building design of high architectural quality.

The scope and content of the tasks was detailed at a previous Contractors' Meeting (1).

2.0 PROGRESS TO DATE

2.1 Consultation

A background paper (2) was commissioned from Loughlin Kealy, an architect and educationalist, to assist in the definition of an educational framework for the remainder of this project and for SOLINFO. This background paper examines the ARCHISOL and SOLINFO action statements as educational initiatives. It discusses issues underlying the task of disseminating solar energy information as outlined in the statements, with particular reference to the audience of architects.

The purpose of the proposals set out in ARCHISOL is to bring about a change in attitudes and behaviour in certain key groups of people concerned with the design and construction of buildings. The fact that one is concerned with bringing about a change in attitudes and behaviour means that one is concerned with education rather than simply with information.

Recognising the importance of building on, rather than duplicating, existing relevant work in this area, it was decided that a preliminary national review task should be carried out in each Member State. The intention was that analysis of the review task reports would assist in identifying institutions and individuals with whom to collaborate.

The authors of the national review tasks (see Appendix I) were requested, inter alia, to:

- Collect examples of nationally-produced introductory and/or promotional passive solar design literature;
- List national organisations of architects;
- List widely-circulated architects' professional journals and magazines;
- List relevant conferences now planned;
- List relevant exhibitions compiled or planned;
- List relevant competitions or awards existing or planned.

Certain of the information thus obtained would also be of value to SOLINFO (3). The reports of the national review tasks were presented in their preliminary versions at an initial project meeting in Dublin on 13 and 14 October 1986. Most of the reports were subsequently added to and furnished in their final form in early 1987. The reports were then analysed (4) in order to clarify the most effective forms of intervention and support.

An expert committee comprising national sub-contractors forms the basis of an information network and is also intended to act as an advisory group which undertakes periodic review of the direction and progress of both ARCHISOL and SOLINFO. There was an informal project meeting in Munich on 9 April 1987 during the European Conference on Architecture. The next project meeting is planned for Porto on 26 and 27 November 1987. Several means of improving communications, both within the group and at regional, national and international level, were considered (5) and a decision has been taken to rely primarily on the system known as EuroKom. Established in 1983 to cater for participants in the European Community's ESPRIT Programme, EuroKom now provides electronic mail and computer conferencing services to industrial and commercial corporations, government organisations and research institutes.

The design of this project and of SOLINFO is such that the administration of many of the component tasks is decentralised, leaving to the centre the primary role of co-ordination and guidance. Reliable and effective communications are crucial to the success of the research. The team of national sub-contractors, currently being completed, provides a channel for administration of some tasks, and a co-ordination and advisory forum for the project as a whole. National sub-contractors are required to possess or develop a wide range of contacts, including editors, journalists, policy-makers in the architectural profession, university and mid-career teachers and instructors, and so on. They are responsible for national dissemination of information, liaison with a wide variety of individuals and organisations, and discussions with selected magazines. They also periodically update the national review tasks carried out at the start of the project.

Fig 1: Project Monitor news release

2.2 Professional Media

A data base on European architectural periodicals has been established. The data base typically contains information such as the name and address of the publiction, the editor's name, the frequency of publication and circulation figures, as well as, in certain cases, an outline of the periodical's editorial policy regarding specific relevant subject areas. Contact has been established with a number of professional journals and magazines, as well as with technical journalists. These initiatives have already resulted in wide coverage of news releases relating to the CEC Project MONITOR brochures (6) and a number of significant reviews of the 1987 European Conference on Architecture (7), particularly in the Federal Republic of Germany, Ireland, the Netherlands and the United Kingdom. Another development which appears to have major potential relates to the MINITEL system, where CEC information dissemination is being conducted by the Comité d'Action Pour le Solaire. MINITEL has in excess of two million subscribers throughout France, of whom thirty five thousand are architects.

Fig 2: Reviews of 1987 European Conference on Architecture

Dialogue has been inititated with the International Association of Architecture Critics (CICA) and this is expected to contribute to increased architectural criticism of selected passive solar buildings. Initially it is proposed to give emphasis to those projects selected for publication by Project MONITOR.

Proposals are being developed for an exhibition and catalogue of European regional vernacular energy-conscious buildings. Proposals for touring the exhibition are also under consideration.

A number of key professionals with responsibilities for design direction, such as chief architects, have been identified and a structure is being developed within which relevant information may be provided to them. A short introductory presentation of the aims of the ARCHISOL and SOLINFO projects was made to the 10th Presidents' Meeting of UIA (Union Internationale des Architectes) Region I (Western Europe) on 19 October 1986 in Paris. The presentation was positively received and the Region I Presidents resolved to be of assistance in disseminating information on the Solar Energy Applications in Buildings sub-programme. This was followed by a 'spontaneous presentation' to delegates at the UIA World Congress in Brighton in July 1987. A data base on European architects' professional organisations has been established.

2.3 Lay Literature

A video programme on passive solar energy for lay audiences is under consideration. The target audience would be those lay people who have an involvement in organising and funding building and renovation projects but who do not themselves have any technical training. With this target audience in mind, the video could also have a wider usage, such as an introduction to mid-career training courses and in schools and colleges. The style of presentation would be popular, with the maximum use of images and diagrams, rather than words, to explain concepts. The programme, approximately 20 minutes in length, would introduce the subject of passive solar building, discuss the concept of climate-responsive architecture, demonstrate the basic theoretical principles in layman's terms, illustrate built examples of different techniques for passive heating and cooling, highlight the potential for energy saving and demonstrate the activities of the CEC in this field. Proposals aimed at second-level education are also being developed.

2.4 Competitions

It is agreed to co-operate with a major Portuguese competition which is planned for 1988. Parallel to the PLEA 88 (Passive and Low Energy Architecture) Conference, there will be a design competition for an energy-efficient passive solar multifamily apartment building on a site at Vila do Conde, 20 km north of Porto.   The competition is due to be announced in the near future, and will be open to all interested architects, with announcements in both Portuguese and English. Specifically, support for the competition will relate to language translation, a workshop on the fundamentals of energy-conscious design and passive solar energy, publicity and an exhibition of entries.   It may also be possible to provide assistance with design file preparation, selection of an international architectural jury and technical assessment of the predicted energy performance of the various competition entries.

It has, nevertheless, proved more difficult than anticipated to obtain useful information regarding relevant forthcoming architectural competitions, especially those competitions which do not already include energy criteria.   Despite their often lengthy lead times, only occasionally is it possible to obtain advance notice of planned competitions, as many regions and countries lack a co-ordinating authority in this area.

Plans for the Third European Passive Solar Competition are at an advanced stage.   This will be a competition for non-domestic passive solar design ideas based on a brief for a new commercial project and

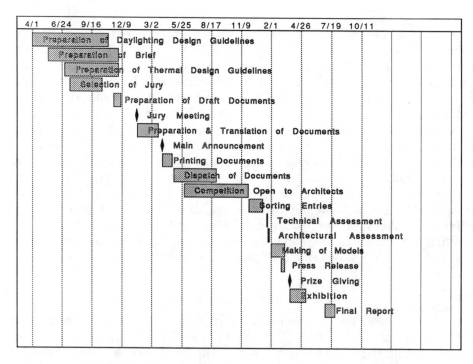

Table I: 3rd CEC Passive Solar Competition: Apr'87 - Jul'89

emphasising natural lighting.   The competition will be announced at Easter 1988 and will close for entries at the end of 1988.   Separate categories for architects and students of architecture are proposed. Several internationally-distinguished European architects have already indicated their agreement to act as judges for the competition and the remainder of the jury will soon be appointed on the basis of an approved shortlist.   The preliminary meeting of the architectural jury for the competition is scheduled for January 1988.

Since 1956, the UIA (Union Internationale des Architectes) has been entrusted by UNESCO with the task of supervising the launching of international competitions in architecture and town planning, and has become the only international consultant on this subject.   On the basis of preliminary draft documents, presented at a meeting in Brighton in July 1987, Gérard Benoit, Director of the UIA International Competitions Programme, has indicated that the UIA would be most pleased to collaborate with the CEC on the Third Passive Solar Competition.   Such collaboration will result in substantial publicity and recognition for the competition.

2.5 Realisation

A series of calculation procedures and performance indices are being developed for use in competitions and in the assessment of design proposals.   A sub-contract has been concluded with Marc Fontoynont, Lyon, for the development of a series of daylighting design guidelines for use in the Third Passive Solar Competition.   Daylighting techniques are passive solar strategies and are particularly attractive for use in buildings with daytime occupancy, such as commercial buildings.   The proposed design methods include evaluation of natural light coming both from the sun and the sky.   The task is scheduled for completion in October 1987.   A proposal for the development of a graphical manual method of assessing the thermal performance of design options is under discussion with Cambridge Architectural Research Ltd.   The method will aim to provide quantitative guidance on as many innovatory features as possible - conservatories and light shelves, for example - but where the performance of such features is not known accurately, cautions will be given.   It is proposed that the method should relate to the approach of the Methode 5000, but attempt to be much simpler in execution.   Both methods will emphasise graphical approaches.

Early in 1987 arrangements were made for the translation into all official languages of the Community (see Appendix II) and dissemination of the Call for Proposals for the new Commission pilot project, BUILDING 2000 (8).   Subsequently, a poster was prepared and presented at the 1987 European Conference on Architecture in Munich. BUILDING 2000 offers support during the design phase for projects in the non-domestic building sector, such as office buildings, hotels, schools and hospitals, which promote innovative use of passive solar technologies.

3.0 CONCLUSIONS

3.1 Present Status and Future Work

Technology transfer, which in the case of passive solar, is of particular importance (there are, for example, over 200,000 architects in Europe), has received little attention heretofore in Europe. However, the necessary objectives, methods and organisation structures for such a project have now been identified. An expert Committee is being established and we are now implementing the priority tasks of:

- Contacting journalists and critics to review selected passive solar buildings;
- Developing proposals aimed at non-technical audiences;
- Arranging workshops, competitions - in particular, the third CEC Passive Solar Competition - and exhibitions.

4.0 REFERENCES

(1) Lewis, J. O., S. O'Toole. Architecture and Solar Energy: ARCHISOL in Solar Energy Applications to Buildings and Solar Radiation Data - Proceedings of the EC Contractors' Meeting held in Brussels, Belgium, 13 and 14 November 1986. D. Reidel Publishing Company, Dordrecht, 1987.

(2) Kealy, L. SOLINFO and ARCHISOL Background Paper: Hidden Agendas. Unpublished paper prepared for Energy Research Group, Univeristy College, Dublin, 1986.

(3) Lewis, J. O., S. O'Toole. Technology Transfer from the Solar R & D Community to the European Building Professions: SOLINFO in Solar Energy Applications to Buildings and Solar Radiation Data - Proceedings of the EC Contractors' Meeting held in Brussels, Belgium, 13 and 14 November 1986. D. Reidel Publishing Company, Dordrecht, 1987.

(4) O'Toole, S. Analysis of SOLINFO and ARCHISOL National Review Tasks. Energy Research Group, University College, Dublin, 1987.

(5) Goulding, J., D. Brickenden. Recommendations for a Communications System for ARCHISOL and SOLINFO. Unpublished report prepared for Energy Research Group, University College, Dublin, 1987.

(6) Turrent, D. Project MONITOR in Solar Energy Applications to Buildings and Solar Radiation Data - Proceedings of the EC Contractors' Meeting held in Brussels, Belgium, 13 and 14 November, 1986. D. Reidel Publishing Company, Dordrecht, 1987.

(7) Palz, W., Editor. CEC 1987 European Conference on Architecture - Proceedings of an International Conference held at Munich, Federal Republic of Germany, 6 - 10 April 1987. H. S. Stephens and Associates, Bedford, 1987.

(8) Commission of the European Communities. Pilot Project: Building 2000 - Call for Proposals. Official Journal of the European Communities, No. S 47/38, 7.3.87.

APPENDIX I: Authors of National Review Tasks

Refer to Appendix I of accompanying paper, 'Technology Transfer from the Solar R & D Community to the European Building Professions: SOLINFO'.

APPENDIX II: Authors of Translations of the Call for Proposals, CEC Pilot Project, BUILDING 2000

- Alex LOHR, Büro für energiegerechtes Bauen, Köln (D).

- Poul LODBERG, Architekt MAA, Aarhus (DK).

- Albert MITJA I SARVISE, Direcció General d'Energia, Generalitat de Catalunya, Barcelona (E).

- P. ACHARD, ENSMP, Valbonne (F).

- A. N. TOMBAZIS, Alexandros N. Tombazis & Associates, Athens (GR).

- Paola FRAGNITO, Progettazione e Ricerca, Milano (I).

- Cees DEN OUDEN, *ARCHI*TECH ABI, Dordrecht (NL).

- Eduardo MALDONADO, Departamento de Engenharia Mecânica, Universidade do Porto (P).

APPENDIX III: Sub-Contractors

- National Sub-Contractor, BELGIUM AND LUXEMBOURG
  André DE HERDE, Centre de Recherches en Architecture, Université Catholique de Louvain.

- National Sub-Contractor, FEDERAL REPUBLIC OF GERMANY
  Alex LOHR, Büro für energiegerechtes Bauen, Köln.

- National Sub-Contractor, GREECE
  A. N. TOMBAZIS, Alexandros N. Tombazis & Associates, Athens.

- National Sub-Contractor, ITALY
  Paolo OLIARO, Instituto di Fisica Tecnica, Politecnico di Torino.

- National Sub-Contractor, PORTUGAL
  Eduardo A. MALDONADO, Gabinete de Fluidos e Calor, Universidade do Porto.

- Sub-Contractor, DAYLIGHTING DESIGN GUIDELINES
  Marc FONTOYNONT, Lyon.

- Sub-Contractor, PASSIVE SOLAR COMPONENTS CATALOGUE
  The ECD Partnership, London.

THE PROJECT PASSYS

Contract number	EN3S-0034-F
Duration	36 months 1 April 1986 – 30 March 1989
Total budget	FF 2 610 000 CEC contribution 100%
Head of Project	Prof. R. GICQUEL Ecole des Mines de Paris, Sophia Antipolis
Contractor	ARMINES
Address	60 Bd. St. Michel F-75272 PARIS CEDEX 06

Summary

The project PASSYS is a european concerted action in the field of passive solar component and system testing.

This paper first presents recent developments in the definition of the different equipments for PASSYS. Those equipments include the test cells, their south walls, the heat removal and heating systems, the instrumentation and the data acquisition system for the cells, and the modelling equipment.

The on-going activities are also presented. They mainly concern test methodologies, model development and validation, simplified design tools and instrumentation. The status of the research in these fields within PASSYS is given.

Introduction

The project PASSYS is a european concerted action in the field of passive solar component and system testing. The PASSYS group is composed of those teams which were selected among applicants to the R/D call for proposals n° C69 of 16 March 1985 : Belgian Centre Scientifique et Technique de la Construction, Deutsche Forschungs- und Versuchsanstalt für Luft- und Raumfahrt, Technical University of Denmark, French Centre Scientifique et Technique du Bâtiment, Italian Centro di Ricerca Termica e Nucleare, Dutch TNO Institute of Applied Physics, and the University of Strathclyde (UK). Two other teams have recently joined the former group, that are the Spanish Instituto de Energias Renovables and the Portuguese Faculdade de Engenharia (University of Porto).

The project PASSYS has been initiated by the Commission of the European Communities in the frame of its Solar Energy R/D Programme. The project is far more than 50% financially supported by the Commission.

The general objectives of PASSYS are the following :

■ increase confidence in passive solar simulation models and design tools which have to be further developed and validated

■ develop reliable and affordable test procedures for passive solar components (PSC) and systems.

A detailed presentation of the PASSYS project was reported at the previous EC Contractor's Meeting in 1986 [1]. It included the rationale for the selection of a test cell for PSC testing and the description of the design of such a test cell and of the different areas of research.

We will now briefly recall the principal steps of the program during the first year.

On the basis of a study by N. Baker, it was decided to concentrate efforts on highly standardized outdoor passive solar test cells, designed to provide a realistic and clearly definable test environment.

A detailed test cell project was prepared by the DFVLR and finally accepted by the PASSYS group.

At the end of 1985, the CEC Joint Research Center (JRC) at ISPRA (I) ordered two such test cells to the german firm Cadolto, selected following a call for bids.

These two cells were delivered in Ispra at the end of April 1986. They were submitted to a preliminary test programme which led to propose some modifications.

An improved test cell design was discussed with Cadolto and 17 units were ordered in 1986 and delivered in the different national test centres in 1987.

Subsequently, up to six test cells per centre will be built.

Decisions concerning the Data Acquisition System for the test cells and the first South components to be used were also taken last year, but the choice of the Heat Removal System was postponed due to technical reasons.

With regards to modelling, on the basis of the previous inter-model comparisons by the Passive Solar Working Group, the Commission has selected the ESP code developed by ABACUS (Glasgow, UK) as the european simulation model for passive systems.

In order to guarantee the full compatibility between modelling and validation activities of the diverse participants, the PASSYS group decided that all teams would purchase the same simulation computer, the MG-1 from Whitechapel (UK), which presents a very good performance / cost ratio.

We will now have a look at the recent developments in the definition of the different equipments for PASSYS. Another chapter will deal with the on-going activities in PASSYS and mainly the research work.

1. Status of the equipments

1.1 Test cells

Some small final modifications to the test cells were requested by the participants in 1986 during the first plenary meeting in Ispra, following the observation of the prototypes built by the Cadolto Company.

Test cell delivery at Sophia Antipolis

Tests carried out on the prototypes having been successful, production started at Cadolto and delivery to all test sites lasted from April to July 1987. Two test cells were delivered to BBRI in Brussels (B), two to TIL in Lyngby (DK), one to CSTB in Sophia Antipolis (F), two to CEA in Cadarache (F), two to Conphoebus in Catania (I), two to DFVLR in Stuttgart (FRG), two to TNO in Delft (NL) and four to University of Strathclyde in Glasgow (UK).

1.2 South walls

The test cells have been delivered jointly with an adiabatic wall and with the reference wall constructed by the Gibat company.

The reference wall is a sandwich panel (concrete / extruded polystyrene / concrete) with a direct gain fenestration (double-glazing). The adiabatic wall contains 40 cm of polystyrene (details on the walls are to be found in [2]).

1.3 Data acquisition system (DAS) for the test cells

The Centre d'Energétique of Ecole des Mines de Paris offered to provide the PASSYS group with a proposal for the hardware specification and with the DAS software [3, 4].

The proposal for the hardware comprises a Hewlett Packard A600 real time computer system, and associated acquisition box HP3852 running under the RTE-A operating system (see figure 1). This is a multi-task, multi-user mini-computer with a 40 Mbyte hard disc and tape back up facilities.

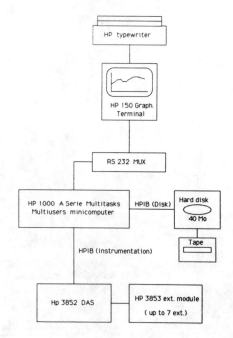

Figure 1 : Hardware DAS configuration [3]

The software developed by Ecole des Mines de Paris is composed of a set of programs allowing for the acquisition of up to 150 channels every basic period (which lowest value is 30 s). Each of the ten possible users can select his proper acquisition period and his proper list of measurements to acquire. The software manages the sharing of the common devices between users so that there is no disturbance from one user to the other. The program has been developed to reach a high level of user-friendliness.

A course on the DAS has been held in Sophia Antipolis from 21st to 23rd April 1987. The hardware was ordered and received by the participants that have test cells and, after some teething troubles, most of them have successfully implemented the software.

1.4 Heat removal system of the test cells

The primary design of the heat removal system (HRS), an airbased system, was rejected by the participants in PASSYS because it would lead to important measurement difficulties.

The new design calls for a liquid (water) based, freeze protected, refrigeration unit providing a high ventilation rate via a fan cooled heat exchanger. The original choice of the Clim 2000 system (F), which had been under test at JRC Ispra along with the test cells, has been rejected due to technical problems (some requirements of the design, e.g. cooling power, could not be reached).

Therefore DFVLR has come up with a new specification [5], (see figure 3). It includes a sophisticated air distribution system to control the air velocity pattern (figure 2).

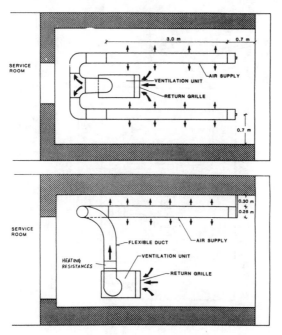

Figure 2 : Vertical section and floor plan
of the air distribution system [5]

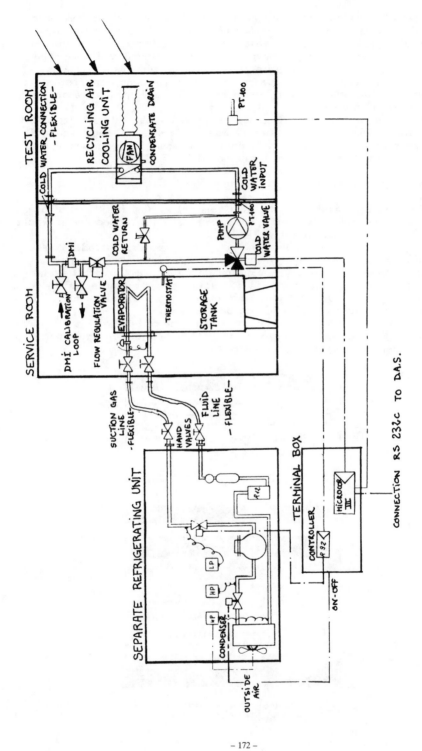

Figure 3 : Schematic layout of the heat removal system [5]

On the 22nd July 1987 the participants decided to choose the Weiß Company (FRG) for the construction of the equipment. Delivery should start in November.

1.5 Heating system and controls

A proposal by M. Chantant (CEA) and D. van Hattem (JRC Ispra), [6], is under investigation at this time.

The heating system proposed consists essentially of a number of heating resistances of different values to be placed in the fancoil unit of the HRS. By switching them on and off in an appropriate way, an almost smooth power modulation can be obtained. The emitted power can be measured without any problem.

A set of three PID controllers will control both the HRS and the heating system, according to the setpoint for the air internal temperature provided by the DAS.

1.6 Modelling equipment

New versions of the Unix operating system (42 nix release 2.5) and of the ESP software have been provided to the participants by Abacus. The new version of ESP includes improvments and new developments.

2. On-going activities

The working structure of PASSYS is based on the concept of subgroups which gather those interested in a common theme.

At the moment four subgroups are activated ; three of them correspond to the main research areas identified for PASSYS : testing methodologies, model development and validation, simplified design tools. A fourth subgroup, the instrumentation one, is specially devoted to the definition and the good use of the equipment. For each of these subgroups, a "leader" has been apointed to take the organization of work in charge.

The progress status is not exactly the same in each of these subgroups. In all of them, the main tasks to be undertaken have been identified and allocated, and basic information is being circulated among participants.

The overall allocation of tasks among participants is shown in Table I.

A short review of on-going activities is given below.

2.1 Test methodologies

One of the main activities of the PASSYS programme is the definition of test methodologies for test cell experiments. By the end of the 3-year programme the group of participants intend to propose one or more experimentally checked test procedures for the evaluation of passive solar components.

The subgroup on test methodologies has been assigned the task of collecting available experience, comparing the different approaches, helping the exchange of results and information among laboratories and producing proposals for the running of test cell experiments. The subgroup leaders are Dick van Dijk, from TNO Delft, and Marcello Antinucci from Conphoebus, Catania.

TABLE I

a: Test cell design
b: Component selection
c: Instrumentation
d: Model development and validation
e: Test methodologies
f: Simplified design tools
g: Experiments

	S		UK				P	NL		I		F			DK	D		B		ASSIGNATION OF TASKS	
	J.R.C. ISPRA	I.E.R.	PILKINGTON	B.R.E.	A.B.A.C.U.S.	E.S.U.	Univ.PORTO	FAGO	T.N.O.	CONPHOEBUS	ARMINES Sophia	ARMINES Paris	C.E.A.	C.S.T.B.	T.I.I.	Univ. Stuttgart	D.F.V.L.R.	V.U.B.	B.B.R.I.		
	CC						FP										CC			ROLE	a
	■										■		■				■			EQUIPEMENT	
	CC		FP										FP				CC			ROLE	b
																				INDUSTRIALS COMP	
																				COMP FOR MODEL	
	CC										CC								LDR	ROLE	c
	■								■											BASIC SCHEME	
																				SPECIFIC TESTS	
			FP	CC											LDR					ROLE	d
											■			■						REVIEW	
														■					■	TECHN.OVERVIEW	
														■						CLIMATE DATA	
				■				■												INT. CONVECTION	
				■																EXT.CONVECTION	
				■																INT. LW RADIATION	
				■																EXT. LW RADIATION	
							■													SHORTW PROCESS	
							■				■								■	TRANS.CONDUCTION	
				■							■									PLANT & CONTROL	
																				P.S.C.	
								■												AIR FLOW PROCES.	
									■											COMFORT	
				■																MODEL ASSESSMENT	
								■	LDR	LDR	CC									ROLE	e
																		■		COMPARATIVE	
	■												■							SUBTRACTIVE	
																				IDENTIFICATION	
											CC			LDR			FP	FP		ROLE	f
									■									■		HEAT REQUIREMENTS	
					■			■												PSC PERFORMANCE	
						■														COMFORT	
																				INTELLIGENT D.T.	
																				PASSYS TEST CELLS	g
	■		■		■		■		■					■						OTHER T.C. DATA	
																				VALIDATION DATA	

– 174 –

A small meeting of experts was held at Conphoebus, Catania, Italy on 27th to 28th November 1986. The aims of the meeting were to review the existing procedures on passive component evaluation in test cells, to identify areas where further work is required, and to develop the draft of a subgroup document on test cell methodologies. At the meeting attention was primarily focussed on a comparative evaluation of the existing test methodologies. A second meeting of experts was held at Sophia Antipolis on 4 March 1987 prior to the PASSYS plenary meeting.

In between, a report on the state-of-the-art in test methodologies was produced [7].

The presentations in Sophia Antipolis, clearly demonstrated the progress made in the previous few months. An important tool in the development of a feasible test methodology is the use of simulated or real test cell data.

The experience thus acquired appears to lead to a significant convergence from originally quite different approaches into three main categories : the comparative method, the absolute one and the model identification techniques.

In the first approach a simultaneous testing in a test cell with the passive solar component (PSC) and in an identical test cell with a reference south façade is performed. The comparison of auxiliary heating of test and reference cells leads directly to the relative performance of test cell with the PSC, compared to the reference cell, as a function of some climate parameter (s).

In the second approach all heat fluxes in the test cell are measured, apart from the flux through the passive solar component subjected to the test. In general the heat fluxes to and from the PSC cannot be measured directly with sufficient accuracy.

The heat fluxes may be measured with short time intervals or, on the contrary, as integrated values over large (e.g. day, month) time intervals. In the first case the heat balance at each timestep (15 to 60 minutes) provides directly the transient net heat flux through the PSC, as function of some climate parameters. A limitation to this method is to get measurements of heat fluxes into the test cell lining. In the second case the large time intervals are needed to avoid the need for measuring the heat flux into the mass of the test cell. An integrated net heat flux through the PSC is obtained, as a function of some climate parameters.

In the model identification techniques a simplified model of test cell and PSC is assumed or created. Computer analysis of the input/output relation provides the parameter values of the model. The "input" consists of measured quantities like solar radiation, outdoor air temperature and auxiliary heating, the "output" is the resulting indoor air temperature.

During that meeting, the different phases of work for the three approaches were elaborated : development of the method, validation against simulation and experimental data, comparison with other test methodologies, and impact on simplified design tools development. A timetable for the different phases and an assignation of tasks were also defined (cf Table I).

During a third meeting of experts in Delft (NL) on 11th and 12th June 1987, several applications of the methodologies to experimental or simulated data were presented.

Topics for futher research are the validity of the methodologies which are being developed. But not only the validity of providing some objective performance characteristics for the PSC itself, but also for the interaction of the PSC with the test cell behind it.

Such interactions are the effectiveness of accumulation and storage of excessive (solar) heat in the cell's mass and the influence of the control of the test cell's auxiliary heating on the utilised solar gains. Other topics that will come under investigation are : the form of presentation of the characteristics, the impact on the development of simplified design tools (e.g. the transferability of results to real buildings), and the evaluation of the effects on results of some "disturbances" such as the effect of wind, of thermal radiation, of occupancy, ...

2.2 Model development and validation

The main aims of the validation subgroup (leader : Lars Olsen from the Thermal Insulation Laboratory of the Technical University of Denmark) is to investigate, principally at the fundamental and algorithmic level, the nature of the physical processus modelled by ESP, and similar programs. To this end the subgroup has reviewed the methodology and theory employed in past validation work, and the next step will be to develop our own working program. This program will concentrate on individual processes within ESP, looking at the consequences of any innaccuracies in the measurement of the input data and how these errors will affect the output parameters.

The subgroup shall also be determining the sensivity of the system to data input/modelling assumptions. Initially it will concentrate on reviewing and understanding the theory, comparisons with experimental results will come later when it is able to recommend the specific measurements and experiments required.

Meetings of the subgroup have been held in Paris (October 86), in Sophia Antipolis (March 87) and in Glasgow (June 87). Presentation at these meetings included reviews of past validation work, the first studies on the topics above evocated and suggestions to improve ESP.

A preliminary technical overview of ESP has been issued by ABACUS and a second course on the software and on the MG-1/ has been given at Glasgow in June 1987.

2.3 Simplified design tools

The simplified design tools (SDT) subgroup was faced with two choices, either to concentrate on the concept of low cost simplified design tools (manual or on micro-computer) or to develop intelligent design tools based on powerful graphical interfaces with sophisticated codes such as ESP.

After discussion with the SOLINFO and ARCHISOL projects manager, J. Owen Lewis, who said professional bodies in Europe clearly indicate their preference for the first approach, it was decided to have the main thrust of the PASSYS work towards low cost SDTs.

The first meeting of this subgroup led by Luc Bourdeau from CSTB (F), was held at Ecole des Mines in Paris on 17 October 1986. Papers on the state of the art in the field of SDT were presented as well as the concept of intelligent design tools in the UK.

At the second meeting at Sophia Antipolis on the 2nd of March, a specification was proposed stating that an ideal SDT should be adaptable to all stages of the architectural design process including the sketch design stage as well as agreed calculation methods for the detailed appraisals that would allow the assessment of PSC performance. The proposal follows the general belief that detailed simulations are not suitable for most users. However the idea that a design tool could be built using models like ESP deriving their simplicity from an intelligent, knowledge based user interface has not been rejected.

Finally, four main topics were identified : heat requirements (heating needs of a building) assessment ; PSC performance assessment ; comfort assessment, and "intelligent" design tools (see also Table I).

2.4 Instrumentation

The aim of the instrumentation subgroup, whose leader is P. Wouters from the CSTC (B), is to define a comprehensive strategy for the deployment and utilization of the equipment so as to ensure that the resulting data is of the most use to the project.

Therefore one of the first tasks of the subgroup was to define a restricted list of instrumentation for the first year of testing [8]. The investigations provided the answers to the following questions : what measures are necessary ? what type of sensors are therefore needed, of which mark, and where to place them ?

A subgroup meeting on instrumentation was held in Sophia Antipolis jointly with the DAS course on April 24th 1987. The topics under discussion included an overview of emergency systems in case of power failure, the nature of two dimensional heat losses in the test cell (important for methodologies and validation), effects of metal plates on the accuracy of temperature and heat flux measurement. Also under review were suggestions concerning the future development of the DAS software and the implementation of common procedures and common equipment for the airtightness measurements of the test cells.

The tasks of the subgroup in the future will include the follow-up of the heating system design (including control) and recommendations for the calibration and use of the equipment (sensors, ...), definition of specific instrumentation schemes for model validation test campaigns, ...

2.5 Miscelleanous

During the second PASSYS plenary meeting held in Sophia Antipolis on the 4, 5 and 6 March 1987, many topics were discussed such as those already presented in this paper (HRS, DAS, instrumentation, ESP, ...).

Some other topics were :
■ the editing of the PASSYS Newsletter whose first issue was released in June 1987 ; the ABACUS team is in charge of this work,
■ the selection of a numbering system for the PASSYS documents,
■ the preparation of a nomenclature and a glossary to be use within PASSYS (ESU team),
■ the format of plenary meetings [9].

3. References

[1] Gicquel, R., "The Project PASSYS", Solar Energy Applications to Buildings and Solar Radiation Data, Proceedings of The EC Contractors' Meeting held in Brussels 13 and 14 November 1986, D. REIDEL, Publ. Comp 1987.

[2] Gicquel, R., Cools, C., "PASSYS Status Report", CENERG, ENSMP, Sophia Antipolis, July 1986.

[3] Mayer, D., "The Data Acquisition Software for the PASSYS Program", CENERG, ENSMP, Sophia Antipolis, April 1987.

[4] Gschwind, M., "Data Acquisition System", CENERG, ENSMP, Sophia Antipolis, April 1987.

[5] Reitz, H.J., Mehlhorn, F., "PASSYS Heat Removal System", DFVLR, FRG, May 1987.

[6] Chantant, M., Van Hattem, D., "Heating and Cooling System for PASSYS Test Cells, A Proposal for Instrumentation and Control", JRC Ispra (I) and CEA Cadarache (F), January 1987.

[7] Van Dijk, H.A.L., "Summary on the State-of-the-Art in the Test Methodologies", TNO Delft, NL, February 1987.

[8] Wouters, P., L'Heureux, D., "Proposal for Instrumentation to be Used by all the PASSYS Teams in the First Measurement Year", BBRI, Belgium, January 1987.

[9] Cools, C., Gicquel, R., "Summary of Decisions Taken During the Second PASSYS Plenary Meeting in Sophia Antipolis", CENERG, ENSMP, Sophia Antipolis, March 1987.

CONTRACTORS' COORDINATION MEETING

Research Areas A&D
1 & 2 October, 1987

LIST OF PARTCIPANTS

NAME   ADDRESS, TEL. & TELEX N°

ADNOT J.   ARMINES
                                        Centre d'Energétique
                                        60, boulevard Saint-Michel
                                        F - 75272 Paris Cedex 06
                                        Tel. (1) 43 29 21 05

ANDREADAKI-CHRONAKI E.   Aristotles University of
                                        Thessaloniki
                                        Faculty of Technologies
                                        Department of Architecture
                                        GR - 54006 Thessaloniki
                                        Tel. (031) 99 15 03, 424-435

ANTINUCCI M.   Conphoebus scrl
                                        Via Leopardi 148
                                        I - 95127 Catania

BACOT P.   NAPAC
                                        25, rue de l'Abbé Groult
                                        F - 75015 Paris
                                        Tel. 48 42 46 74

BERGMEIJER P.W.   Technisch Physische Dienst
                                        TNO-TH
                                        P.O. Box 155
                                        NL - 2600 AD DELFT
                                        Tel. (015) 78 80 20

BOFFA C.   Politecnico Dip. Energetica
                                        Corso Duca Abruzzi 24
                                        I - 10129 Torino
                                        Tel. (11) 54 78 59
                                        Telefax (11) 54 78 59
                                        Telex 220646 POLITO I

BOURDEAU L.   Centre Scientifique et Technique
                                        du Bâtiment
                                        BP 21
                                        F - 06561 Valbonne Cedex
                                        Tel. 93.65.34.00
                                        Telex 970194 F

BOURGES B.	6, rue de l'Armor F - 35760 St-Grégoire Tel. 99 38 94 47 Telex 950 198 F ITARE
BOYEN H.	Koninklijk Meteo Instituut Ringlaan 3 B - 1180 Brussel Tel. (02)375 24 78 Telex 21315
CARABATEAS E.N.	Secretariat General Research & Technology Messogion 14-18 GR - Athens Tel. 771 15 19
CASTRO DIEZ Y.	Dpto. Fisica Aplicada Facultad de Ciencias Universidad de Granada E - 18071 Granada
CHAUVEL P.	C.S.T.B. Division Eclairage 11, rue Henri-Picherit F - 44300 Nantes Tel. 40 59 42 55
COCH H.	Escola d'Arquitectura de Barcelona Av. Diagonal, 649 E - 08028 Barcelona Tel. (93) 334 69 95
COOLS Ch.	Ecole des Mines de Paris Centre d'Energétique Sophia Antipolis F - 06560 Valbonne Tel. 93 95 75 75
DEBOSSCHER A.	Katholieke Universiteit Leuven Celestijnenlaan 300 A B - 3030 Heverlee Tel. (016) 22 09 31
den OUDEN C.	Architect P.O. Box 1042 NL - 3300 BA Dordrecht Tel. (31) 78 145 822 Telex 29322 eqm nl
DEPECKER P.	INSA de Lyon LEH Ave Albert Einstein, 20 F - 69 Villeurbane Tel. 78.94.81.12. p 8817

DOGNIAUX R.	148, rue Groeselenberg B - 1180 Bruxelles Tel. (2) 374 34 03
DUPAGNE A.	Université de Liège LEMA 15, avenue des Tilleuls B - 4000 Liège Tel. (41) 52 01 80
DUTRE W.D.	Katholieke Universiteit Leuven Celestijnenlaan 300 A B - 3030 Heverlee Tel. (016) 22 09 31
EID M.	Polytechnic of Central London 35 Marylebone Road GB - London NW1 5LS
FANCHIOTTI A.	Dip. TECA University of Rome Via Eudossiana 18 I - 00184 Roma Tel. (6) 461 045
FANGER P.O.	Technical University of Denmark Lab. of Heating & Airconditioning DK - 2800 Lyngby Tel. (1) 88 46 22 Telex 37529
FURBO S.	Thermal Insulation Laboratory Technical University of Denmark Building 118 DK - 2800 Lyngby Tel. (452) 88 35 11 Telex 37529 DTH DIA DK
GEOGHEGAN P.	Housing and Urban Design Research Unit School of Architecture UCD Richview, Clonskeagh Rd. IRL - Dublin 14 Tel. 01 69 71 11
GRANT M.	ABACUS Dept. of Architecture Strathclyde University GB - Glasgow G4 0NG
GICQUEL R.	Ecole des Mines Sophia Antipolis F - 06560 Valbonne Tel. (93) 95.75.75

GRIFFITHS I.

Dept. of Psychology
University of Surrey
GB - Guildford GU2 5XH
Tel. (0483) 509 267

HUG G.

ADURE/ENSAIS
24, boulevard de la Victoire
F - 67000 Strasbourg
Tel. 88 35 55 05

JENSEN S.O.

Thermal Insulation Laboratory
Technical University of Denmark
Building 118
DK - 2800 Lyngby
Tel. (452) 88 35 11
Telex 37529 DTH DIA DK

JIMENEZ J.I.

Dpto. Fisica Aplicada
Facultad de Ciencias
Universidad de Granada
E - 18071 Granada
(58) 27 28 87

JOUKOFF A.

Institut Royal Météorologique
Avenue Circulaire, 3
B - 1180 Bruxelles
Tel. (2) 375.24.78
Telex 21 315

LEWIS J.O.

University College Dublin
Energy Research Group
Richview, Clonskeagh
IRL - Dublin 14
Tel. (01) 69 71 11
Telex 91178 UCD EI
Telefax (01) 83 89 08

LINDEN J.

Technisch Physische Dienst
TNO-TH
P.O. Box 155
NL - 2600 AD Delft
Tel. (015) 78 80 20

LOHR A.

Oldenburger Str. 68
D - 5000 Köln 60
Tel. (221) 740 77 63

MALDONADO E.

Dept. Eng. Mecânica
Fac. Engenharia
R. Bragas
P - 4099 Porto Codex
Tel. (02) 27505 Ext. 374
Telex 27323 FEUP P

MICHAIL Y.

Ethnoktimatiki spa
Lekka 23
GR - 14562 Athens
Tel. 32 22 577

OLSEN L.

Thermal Insulation Laboratory
Technical University of Denmark
Building 118
DK - 2800 Lyngby
Tel. (452) 88 35 11
Telex 37529 DTH DIA DK

O'TOOLE S.

Energy Research Group
University College Dublin
Richview, Clonskeagh
IRL - Dublin 14
Tel. (01) 69 71 11
Telex 32693 UCD EI
Telefax (01) 83 89 08

PAGE J.

15 Brincliffe Gardens
GB - Sheffield S11
Tel. (742) 55 15 70

PALZ W.

CEC- DG XII
200, rue de la Loi
B - 1049 Bruxelles
Tel. (2) 235 69 22
Telex 21877 COMEU B
Telefax (2) 235 01 45

PARRINI F.

ENEL-CRTN
Via Rubettino 54
I - Milano

RAOUST M.

DIALOGIC
70, boulevard de Magenta
F - 75010 Paris
Tel. 42 06 53 20

RASCHKE E.

Institut für Geophysik u.
Meteorologie
Universität zu Köln
Kerpener Str. 13
D - 5000 Köln 41
Tel. (221) 470 36 82
Telex 888 2279

REITZ H.J.

DFVLR, EN-TT
Pfaffenwaldring 38-40
D - 7000 Stuttgart 80
Tel. (711) 68 62 421
Telex 7255689

SACRE Ch.

CSTB
Division Climatologie
11, rue Henri-Picherit
F - 44300 Nantes
Tel. 40 59 42 55

SCHARMER K.

GET
Gesellschaft für Entwicklungs-
technologie mbH
Industriestr.
D - 5173 Aldenhoven
Tel. (02464) 5810
Telex 833569 budi d
Fax (02464) 58136

SERRA R.

Escola d'Arquitecture de Barcelona
Av. Diagonal, 649
E - 08028 Barcelona
Tel. (93) 334 69 95

SOHNS J.

Inst. für Thermodynamik und
Warmetechnik
Universität Stuttgart
Pfaffenwaldring 6
D - 7000 Stuttgart 80
Tel. (711) 685 35 36

SPENCER H.B.

Scottish Centre of Agricultural
Engineering
Bush Estate
Penicuik
GB - Midlothian, Scotland EH26 0PA
Tel. (31) 445 21 47
Telex 265 871 REF 50E 001

STEEMERS T.C.

CEC - DG XII
Wetstraat 200
B - 1049 Brussel
Tel. (2) 235 68 78
Telex 21 877 COMEU B
Telefax (2) 235 01 45

TRAISNEL J.P.

Ecole Nationale Supérieure
des Mines de Paris
Centre d'Energétique
60, boulevard St-Michel
F - 75272 Paris Cedex 06
Tel. (1) 43 29 21 05

TRYSSENAAR A.

PBE
NL - Bloemendaal

TURRENT D.

The ECD Partnership
11-15 Emerald St.
GB - London WC1N 3QL
Tel. (01) 405 31 21
Telex 261507 Ref. 2924

TWIDELL J.

Energy Studies Unit
University of Strathclyde
GB - Glasgow G1 1XQ
Tel. (041) 552 44 00 ext. 3307

VALKO P.

Swiss Meteorological Institute
Priv. address :
Langwattstr. 26
CH - 8125 Zollikerberg
Tel. (01) 391 75 53
or   (01) 256 93 71

van DIJK H.A.L.

TNO Institute of Applied Physics
P.O. Box 155
NL - 2600 AD Delft

van PAASSEN A.H.C.

Mekelweg 2
TU Delft
NL - 2615 Delft

VEST HANSEN T.

Danish Solar Energy
Testing Laboratory
Technological Institute
Postbox 141
DK - 2630 Taastrup
Tel. 45 2996611
Telex 33416 ti dk

WALKER J.

Polytechnic of Central London
35 Marylebone Road
GB - London NW1 5LS

WELSCHEN B.A.M.

Fugro Geodesic BV
Postbus 1213
NL - 2260 BE Leidschendam
Tel. (070) 11 11 60
Telex 33292 fci nl
Telefax (070) 20 27 03

WILLBOLD-LOHR G.

RWTH Aachen
Lehrstuhl f. Baukonstruktion II
Schinkelstr. 1
D - 5100 Aachen
Tel. (241) 80 38 94

WOUTERS P.

Belgian Building Research
Institute (CSTC/WTCB)
Lombardstraat 41
B - 1000 Brussel
Tel. (2) 653 88 01
Telex 25 682 CELEX B

ZARAMELLA G.

Fiat Engineering
V. Lagrange 7
I - Torino
Tel. (11) 54 66 01